MW00508850

ENTREPRENEURSHIP + INNOVATION IN EGYPT

ENTREPRENEURSHIP + INNOVATION IN EGYPT

Edited by Nagla Rizk
and Hassan Azzazy

The American University in Cairo Press

Cairo New York

First published in 2016 by
The American University in Cairo Press
113 Sharia Kasr el Aini, Cairo, Egypt
420 Fifth Avenue, New York, NY 10018
www.aucpress.com

Exclusive distribution outside Egypt and North America by I.B.Tauris & Co Ltd.,
6 Salem Road, London, W4 2BU

Dar el Kutub No. 2576/14
ISBN 978 977 416 727 0

Dar el Kutub Cataloging-in-Publication Data

Rizk, Nagla
        Entrepreneurship and Innovation in Egypt / Nagla Rizk and Hassan Azzazy.—Cairo:
The American University in Cairo Press, 2015.
        p.   cm.
        ISBN 978 977 416 727 0
        1. Entrepreneurship—Egypt
        2. Azzazy, Hassan (Jt. auth)
        3. Title
        338.040962

1 2 3 4 5   19 18 17 16 15

Designed by Jon W. Stoy
Printed in Egypt

Contents

Tables

Figures

Contributors

Hassan Azzazy is a tenured professor of chemistry at the American University in Cairo. He is the founder of the Novel Diagnostics & Therapeutics Research group. He has served as the chairman of the chemistry department, the founding director of the MSc chemistry program, and the associate dean for Graduate Studies and Research. Before joining AUC, Dr. Azzazy was a postdoctoral fellow and assistant professor at the University of Maryland School of Medicine. He has over 24 years of biomedical research experience. Dr. Azzazy has over 140 scientific publications, all in international refereed journals, conferences, and book chapters. Dr. Azzazy is a strong advocate of entrepreneurial education and has been working with the European Training Foundation to promote the introduction of entrepreneurial learning in higher education in Egypt. Dr. Azzazy established the first Egyptian spinoff company, D-Kimia, LLC, which is focused on the development of innovative diagnostic assays. In 2014, Dr. Azzazy received the Young Innovator Award from Burayda Colleges in Saudi Arabia and the Global Innovator Award from Texas Christian University.

Karim Badr is a financial economist at the World Bank Cairo office in the Finance and Markets unit. He joined the World Bank in January 2010 as an economic research analyst in the Poverty Reduction and Economic Management unit. He has an MBA degree from Maastricht School of

Management, specialized in finance and banking, and a B.A. in economics from the American University in Cairo. Prior to the World Bank, Karim occupied several positions in the fields of banking, investment, and research. He also taught several courses in the fields of economics and finance in several universities in Egypt. He has authored and co-authored several publications in the fields of entrepreneurship and innovation, inequality, and labor markets.

Shima Barakat is an entrepreneur, director, and academic obsessed with making the world a better place. She has spent two decades helping companies, governments, and international funding agencies improve their performance in an environmentally and socially sensitive manner. Shima is the director of two enterprise and entrepreneurship programs for PhDs at the University of Cambridge, UK, supporting the development of technology entrepreneurs and the commercialization of technology from within the university and its partners. As an entrepreneur, Shima is one of the founders and a director of Value in Enterprise, a responsible business consultancy company. She was one the founders of Nahdet El Mahrousa and the Egyptian Junior Business Association (EJB) in Egypt and the Global Communities Initiative (GCI) in the United States, as board chair. Shima is interested in critically studying entrepreneurship practice to explore the implications on people and the planet. Currently, she has a particular interest in gender influences.

Adel Boseli has more than ten years of experience in the software industry with a proven track record in software design, implementation, systems architecture, data analysis, and project management. He is an entrepreneur and co-founder of Shekra Crowdfunding, an Egypt-based company focusing on the entrepreneurial ecosystem in the MENA region and offering sharia-compliant crowdfunding. He is a consultant for the World Bank Group, focusing on SME financing and its models. Adel started his software career after graduating from Ain Shams University in computer engineering, and then successfully delivered a series of software projects around the globe for local and global companies, regional organizations, and governments. In 2007, he started his entrepreneurial path by joining IdealRatings, a San Francisco-based startup, which has grown to be the global leader in Islamic fund management and index services. He has cofounded a number of startups in various industries: medical, real estate, and digital advertising.

Nagwa Ibrahim was adjunct associate professor of practice, Promoting Entrepreneurship, in the AUC School of Sciences and Engineering, and associate provost for development until 2013. She is also the international adviser, director, and a senior trainer for entrepreneurship programs, as well as a member of the International Goldman Sachs Program for 10,000 Women Entrepreneurs in the MENA region. Over the past twenty years, she has been in charge of developing communication strategies and has worked as a senior consultant with Egyptian ministries, TEVT, ILO, the EU, and USAID. Dr. Ibrahim has provided several training courses/programs in career development, leadership and managerial skills/communications, human resources strategies and policies, job analysis, motivation, organizational goals, and recruitment. She is an international adviser for executive education with several top universities in the United States and in the Arab region. She has published several papers on entrepreneurship in top international journals. She was a founding member of the Global Entrepreneurship Monitor (GEM) Egyptian team and of the Middle East Council for Small Business and Entrepreneurship, an affiliate of the International Council for Small Business.

Ayman Ismail is an assistant professor and holds the Abdul Latif Jameel Endowed Chair of Entrepreneurship at the American University in Cairo's School of Business, where he also leads the AUC Venture Lab. Prior to that, he was a consultant with McKinsey & Company, based in New York, and a cofounder and managing partner at Enovio Consulting. He has consulted for international organizations such as the World Bank, IFC, UNCTAD, and several US government agencies. Ayman is active in civil society and in policy advising in Egypt. In 2012, he was selected as a Young Global Leader by the World Economic Forum. He is a cofounder and board member of Nahdet El Mahrousa NGO. He is a former research fellow at Harvard University. He received his PhD in international economic development from MIT. He also holds a bachelor's degree in engineering, an MBA from AUC, and a master's degree from MIT.

Khaled Ismail holds the Willard W. Brown Chair at the American University in Cairo for the academic year 2014–15 and is the head of the Entrepreneurship and Innovation Program (EIP) at AUC. He is the founder and chairman of KIangel, an angel investment fund focused on investing in early-stage startups in the Middle East. He has started seven companies in Egypt since 1992; Intel acquired the latest company, SySDSoft, in

March 2011. Dr. Ismail was the chairman of Endeavor Egypt, which is a global not-for-profit organization supporting high-impact entrepreneurs in developing countries. He has served on the boards of Orascom Telecom and other companies. Dr. Ismail received his PhD from MIT in 1989. He is an IEEE Fellow, has published over 160 papers in international journals, and holds 22 US patents. He was the recipient of the Eta Kappa Nu "Best Young Electrical Engineer in the US" award in 1994, and the Shuman Award for "Young Arab Engineer" in 1995.

Sherif Kamel is professor of MIS and founding dean of the School of Business at the American University in Cairo. He established the Center for Entrepreneurship and Innovation and the Venture Lab, making the university the leading educational partner in the entrepreneurship ecosystem in Egypt. His research and teaching interests include IT management, IT transfer to developing nations, e-business, and DSS. His work is widely published in IS and management journals. He is the editor of three books: *E-Strategies for Technological Diffusion and Adoption: National ICT Approaches for Socioeconomic Development*; *Electronic Business in Developing Countries: Challenges and Opportunities*; and *Managing Globally with Information Technology*. He is the associate editor of the *Journal of IT for Development* and the *Journal of Cases on Information Technology*. Kamel holds a PhD in IS from the London School of Economics, an MBA from AUC, a B.A. in business administration, and an M.A. in Islamic art and architecture from the American University in Cairo.

David A. Kirby is vice president (Enterprise and Community Service) at The British University in Egypt, which he joined in 2007. He has some forty years' experience in entrepreneurship and small business management, and was a pioneer of entrepreneurship education in the United Kingdom, for which he was awarded, in 2006, the Queen's Award for Enterprise Promotion. He is a former director of the UK Institute for Small Business and Entrepreneurship, and a former senior vice president and director of the International Council for Small Business. In recognition of his consultancy and training work with small businesses, he was elected to a fellowship of the Institute of Business Advisers. He has published 150 journal articles and 18 books and research monographs, including *Entrepreneurship* (McGraw-Hill, 2003). His research is currently focusing on entrepreneurship education and innovation and entrepreneurial universities, and he is the research leader for the Global Entrepreneurship Monitor (GEM) Project in Egypt.

Zuhayr Mikdashi is professor emeritus at the University of Lausanne, Switzerland, where he founded the Institute of Banking and Financial Management. He has taught in the areas of banking, and financial systems, and energy economics. He is a non-executive member of the board of directors of a Swiss bank, a French machine tool manufacturer, and three subcontracting industrial companies in the Swiss watch-making sector, assuming the chairmanship of their boards. Other academic-cum-research and advisory functions include those of visiting faculty at the Graduate School of Business, Indiana University; distinguished visiting professor at the AUC School of Business; research fellow at the Center for International Affairs, Harvard University; research scholar at the Japanese Institute for Developing Economies, Tokyo; and visiting scholar at the foundation Resources for the Future, Washington, D.C. He has worked as a consultant for a number of agencies such as the World Bank; as resident advisor to the minister of finance and oil, Kuwait; and as advisor to the World Economic Forum. He earned his PhD in international economic relations from Oxford University.

Andy Penaluna is director of the International Institute for Creative Entrepreneurial Development and holds a Personal Chair at the University of Wales Trinity Saint David in Swansea, Wales. He is a distinguished visiting professor at AUC, a visiting professor at the University of Leeds, and an Innovation Fellow of the Royal College of Art. Andy set up and was the chair of the UK Higher Education Academy's Special Interest Group in Entrepreneurial Learning. He chairs the UK's Quality Assurance Agency's Graduate Enterprise and Entrepreneurship Group. He helped to lead a Welsh consortium of universities and colleges to develop what is believed to be the UK's first fully university-accredited M Level module in initial teacher training for enterprise educators. He is now an executive director developing international dialogue and has contributed to the development of a new international policy toolkit at the UN in Geneva. In November 2014, Andy was awarded the enterprise educator award by the UK sector skills organization for small business mentoring (SFEDI). Andy is also the recipient of the Queen's Award for Enterprise Promotion 2015.

Kathryn Penaluna is the enterprise manager at the University of Wales Trinity Saint David (formerly Swansea Metropolitan University). Her role includes helping graduate businesses to grow and develop, for which she has an enviable track record. Her approaches are driven by experiences of working within both business school and art and design departments, and now she is an advocate of a 'design thinking' stance. Kathryn helped to run the UK Higher Education Academy's Entrepreneurial Learning Special Interest Group and has contributed to educational developments within the Academy's Business Management Accountancy and Finance Subject Centre. She has also worked with the Welsh government as an Enterprise Champion. Her MBA focused on education in the creative industries and opening up new avenues for exploration, and led to working with the UK Intellectual Property Office to develop new teaching strategies. Intellectual property forms an integral part of Kathryn's research, as she focuses on the nexus of business and creativity within enterprise education.

Nagla Rizk is professor of economics and founding director of the Access to Knowledge for Development Center (A2K4D) at the School of Business, AUC. She is a faculty associate at the Berkman Center for Internet and Society at Harvard University and an affiliated fellow of the Information Society Project at Yale Law School. Her area of research is the economics of knowledge, information technology, and development, focusing on business models in the digital economy, intellectual property, and human development. Rizk is a member of the Executive Committee of the International Economic Association, a founding member of the Access to Knowledge Global Academy, and a member of the steering committee of the Open African Innovation Research Project. She wrote the National Strategy for Free and Open Source Software in Egypt. At AUC, Rizk served as associate dean for graduate studies and research in the School of Business and as chair of the economics department.

Shailendra Vyakarnam worked in industry for several years before completing his MBA and PhD. He has combined academic, practitioner, and policy interests to provide advice to government agencies and UN agencies in several countries on the development of entrepreneurial ecosystems, technology commercialization, and entrepreneurship education. He has mentored entrepreneurs and held non-executive directorships of small firms in addition to developing growth programs for SMEs over several years. His main contribution over the past ten years has been to develop practitioner-led education for entrepreneurship at the University of Cambridge Judge Business School, Centre for Entrepreneurial Learning. He has been assisting universities in several countries to better understand how to integrate this novel curriculum into their programs. Dr. Vyakarnam is presently co-founder and director of AcceleratorIndia.

Sherif Yehia has diversified business operational experience in multinational organizations. He previously worked as a network engineer at Orange Business Services and as a process management consultant at HSBC and is currently a supply chain manager at Procter & Gamble. Sherif is passionate about combining his practical experience with research work. He is currently a research assistant at the American University in Cairo, with research interests that include entrepreneurship ecosystem enablers, business model innovation, and supply chain practices in "bottom of the pyramid" markets. Sherif holds a bachelor's degree in biomedical engineering from Cairo University and an MBA from the American University in Cairo, where he received the Jameel Fellowship.

Acknowledgments

We are grateful to our authors for their contributions as well as their patience with our editorial process. We are also thankful for the valuable support of our home institution, the American University in Cairo (AUC), and our respective schools: the School of Business and the School of Sciences and Engineering. This work would not have been possible without the financial support of AUC's Office of the Vice Provost for Research and the AUC School of Business, through both the Office of the Associate Dean for Graduate Studies and Research and the Access to Knowledge for Development Center (A2K4D). Special thanks are due to Sherif Kamel, former dean of the School of Business, for his passionate support of this project since its inception.

We are also grateful to Lina Attalah and the team of *Mada Masr* online newspaper for contributing to the editorial process of this book. The research team at the Access to Knowledge for Development Center has also provided invaluable contributions to the editorial process, and to them we are grateful. Last but not least, thanks are due to the research associates and young scholars who assisted with this project: Nagham El Houssamy, Stefanie Felsberger, Sylvia Zaky, Tamer Fakhereldin, and Yousra Bakr.

Foreword

Sherif Kamel

It gives me great pleasure to see the publishing of this book, which addresses the notion of entrepreneurship and innovation, one of the most influential topics in recent times when it comes to the effective platform for productivity and economic growth in different societies around the world. Entrepreneurship and innovation are here to stay. Their momentum and implications worldwide will increase with the growing emergence and universal access of technology tools and applications, mobility, and interconnectivity that are becoming integral elements of our lives. Entrepreneurship and innovation clearly have the potential to change economies, transform societies, and consequently represent a unique and much needed opportunity to create jobs and improve standards of living, especially in emerging economies. In this pivotal time for Egypt and the Middle East region, the notion of startups, a strong entrepreneurial culture, and an innovative mindset are now more important than ever and are needed to become the driver and catalyst to rebuild Egypt and the region on strong, solid, and sustainable foundations.

Entrepreneurship is not new to Egypt or to the Middle East. Throughout history, Egyptians have been known as successful entrepreneurs across different sectors, moving between provinces in Egypt and across nations in the region, and being actively involved in trading and growing businesses in different sectors. That somehow changed some time ago, following which the aspiration of many Egyptians has been to work for the government and the public sector. The motive was to secure a job with minimal risk and few daily challenges. It became a culture that relied primarily on securing a safe working opportunity regardless of the potential that presented itself elsewhere in the marketplace. This has gradually started to change since the late 1990s, however, with a growing young population that is technology savvy, better educated, more exposed, and willing to venture into the business world at a younger age. In 2008, such change started to take better shape with the proliferation of business associations, organizations, and business plan competitions supported by investors, mentors, local companies, and multinationals. Consequently, in the Middle East over the last six years, more than 140 organizations have been established and/or started to provide different types of support, whether financial or non-financial, to the entrepreneurial ecosystem. Several factors have contributed to such change, including an average population growth rate of 2.1 percent per year in Egypt, whose 85 million people are overwhelmingly young with 58 percent under the age of 25, coupled with a growing belief that the nation's future can only be improved with a more agile and competitive private sector. The change was boosted by the growing diffusion of information and communication technology usage and an increasing investment in entrepreneurial awareness campaigns and educational and training programs.

Accordingly, over the last decade many stories emerged in the Middle East of promising entrepreneurs who have great ideas for startups that can have positive implications for the societies of the region. Given the demographics of the Middle East and with a growing and young population increasingly exposed through various technology and social media platforms, there is no shortage of ideas that can spawn startups in different sectors and industries such as health, environment, tourism, education, agriculture, energy, recycling, music, entertainment, and more. I have been personally privileged to have worked with many of these young entrepreneurs, mentored some, advised others, exchanged ideas with many, and had multiple discussions and panel debates over the future of the region. In my opinion, the future of the Middle East will rely on how effectively the economies in the region will benefit from the power of entrepreneurship. In Egypt, it will rely also on the extent to which the country's invaluable resource for the twenty-first century—people, will help transform the economy in different directions while addressing the multiple prospects across the nation that are mostly untapped.

Since 2008, the School of Business at the American University in Cairo has been promoting the notion of entrepreneurship and innovation on and off campus. Many activities and services have been offered to a growing population of entrepreneurs from across the country. Over the years, multiple events have been organized, seminars and conferences convened, training sessions conducted, and boot camps and business plans initiated and/or hosted with the movers and shakers in the entrepreneurial ecosystem in Egypt and globally. In 2014, the Center for Entrepreneurship and Innovation at the School of Business, which had been a program since 2010, was established to invest in Egypt's promising young entrepreneurs through supporting startups, business plan competitions, boot camps, and mentorship, becoming the country's leading educational partner in the entrepreneurship ecosystem. In addition, the university established the Venture Lab in 2013, which became Egypt's primary university-based incubator focusing on startups that can realize scalable impact on the community.

As the primary entrepreneurial education partner in the ecosystem in Egypt and a strong advocate of the interdisciplinary approach to promoting entrepreneurship and innovation, the American University in Cairo, through the collaboration of the School of Business and the School of Sciences and Engineering, convened a research conference in 2012 called "Entrepreneurship and Innovation: Shaping the Future of Egypt." Multiple presentations and panel discussions addressed various issues and ideas explored by academics and practitioners from different fields and disciplines on the possibly invaluable role that the development of an entrepreneurial culture could have in shaping Egypt's economy and future. This book gathers and documents many of the topics and discussions that took place during the conference into eleven interesting chapters authored by experienced academics and practitioners in the domain of entrepreneurship. They address various issues of primary importance to entrepreneurship and innovation in terms of historical background; policy development; education and lifelong learning; creativity elements; institutional behaviors; the role of academic and institutional settings in promoting the culture of entrepreneurship; the importance, availability, and accessibility of timely and relevant information; entrepreneurs as agents of change and their impact on societal transformation; community stakeholders and the importance of the ecosystem; business models for funding and the associated pros and cons; and more.

The articles and cases represent a bridge between markets, industries, businesses, and academia, and reflect a rich repository of issues, prospects, challenges, and opportunities in entrepreneurship and innovation that could have sustainable and scalable impact on emerging economies, with a focus on Egypt. Entrepreneurship and innovation are transforming individuals, organizations, and societies, and I am confident that this book will have an effective contribution to exploring and promoting these influential trends in society as we move forward. I would like to seize the opportunity and thank all those who were involved in the development and production of this book as well as all the authors, reviewers, and editors whose expertise helped in delivering such an important contribution.

It is my hope that this book will provide emerging economies, academics, and practitioners of entrepreneurship and innovation with a variety of research and practice-oriented views on development, policy, and regulatory and market issues in such an evolving, dynamic, complex, and constantly changing entrepreneurial global marketplace. Eventually, sharing the lessons learned and exchanging all accumulated capacities and knowledge will allow emerging economies and their promising entrepreneurs to leverage their entrepreneurial skills and to benefit from the opportunities constantly emerging in this global, interconnected, and dynamic marketplace.

With a focus on youth, emerging technologies, creative minds, and untapped opportunities, and while addressing the issues of responsible business, leadership, governance, ethics, education, and vocational training—as well as the formulation and institutionalization of the proper regulatory framework and required policies—the creation of a startup mindset will help develop a startup community and consequently a startup culture. I am happy and encouraged that this book sheds light on and addresses many issues that are invaluable in paving the way for the creation of an entrepreneurial culture in the Middle East. More importantly in many ways, this book celebrates the notion of entrepreneurship and, rightly so, focuses on its most important element and primary building block: people.

Introduction

Nagla Rizk and Hassan Azzazy

Egypt has been living a new era since the January 2011 revolution. Of all the challenges that the country has been facing, the economy remains the most pressing. Egypt's GDP growth rate remains at a stagnant 2.3 percent. This is mainly due to the challenges since 2011—namely, a sharply declining tourism sector characterized by low productivity, ripple effects of the global financial crisis, political instability, and insecurity.[1] Investment and financial freedom have been further curtailed amid the economic stagnation with outstanding government debt comprising 89 percent of the country's GDP in 2013, and a budget deficit accounting for almost 12 percent of GDP as of 2014.[2] With recent fiscal reforms and the inflow of grants from neighboring countries, some improvement is expected; However, with a projected economic growth rate of 3.3 percent in fiscal year 2015 and a budget deficit decline of 2 percent, comprising 10 percent of GDP.[3]

At this pivotal juncture, the country is in dire need of novel approaches to boost the domestic economy and advance its utilization of resources. In this context, entrepreneurship and innovation emerge as significant drivers for the country's development on strong, sold, and sustainable foundations. As such, an entrepreneurial spirit and innovative mindset need to be capitalized on, fostered, and encouraged. This takes place against a wealth of human capital, where 23.7 percent of Egypt's population (approximately 20 million) falls in the 18–29 age bracket.[4]

In parallel to fostering entrepreneurship and innovation, there is an urgent need to reform education to encourage students to innovate and to establish national strategies to harness, incubate, and develop innovative ideas in different disciplines. Recently, Egypt has witnessed many activities to support entrepreneurship and innovation in universities. These include the establishment of technology transfer offices and accelerators/incubators in several national and private universities, as well as the organization of business plan competitions and scientific innovation competitions. These aim to offer mentorship and assess inventions and prototypes of relevance to national priorities. The scientific community has become more aware of the significance of intellectual property protection and the need for applied and translational research to address national and global challenges in the areas of food, health, energy, water, pollution, and climate change. Moreover, NGOs have recognized the significance of entrepreneurship and innovation and have dedicated funds to support innovative projects that address social challenges. Although there are some existing technology clusters in Egypt (for example, in the area of information and communication technologies), additional clusters in biotechnology, nanotechnology, and possibly energy should be established.

A few attempts have been made to highlight the significance of the link between industry and academia, including the scouting of innovative solutions in academic departments and bringing industrial challenges to the attention of researchers. In 2010, the American University in Cairo (AUC) launched the Entrepreneurship and Innovation Program, later to be named the Center for Entrepreneurship and Innovation. The university also established the AUC Venture Lab in 2013. The first spinoff nanobiotechnology company (D-Kimia, LLC) to be generated from an Egyptian university was established by AUC in 2013. The Technology Transfer Office at AUC negotiated patent licensing to D-Kimia. The company was successful in raising funds and received support from the Egyptian diaspora.

Acting on its firm belief in the role of entrepreneurship and innovation as drivers for development and in the instrumental role that universities can play in this context, AUC held its annual research conference in April 2012 under the title "Entrepreneurship and Innovation: Shaping the Future of Egypt." The conference was held with the purpose of fostering synergies between entrepreneurship, enterprise development, job creation, and the spread of social benefits in Egypt. It provided a forum for promoting links between entrepreneurs from within the AUC community on the one hand, and business profesionals and regional and global communities on the other.

The conference also introduced a mechanism for shaping the entrepreneurial mindset and the required skill set among younger generations. It served to enhance confidence and cement trust among different categories of investors in Egyptian enterprises. A key theme of the conference was to focus on the development of certain technology sectors. Namely, information and communication technology, biotechnology, renewable energy, and green technologies, all of which bolster the knowledge-based component of Egypt's economy. The conference also included a discussion and identification of the 'best practices' of commercializing and socializing innovation for the purposes of sustainable development.

This book is a collection of chapters based on selected presentations at the conference. This compilation of deliberations is intended to document contributions on key issues, including but not limited to entrepreneurial education, social entrepreneurship, the entrepreneurial ecosystem, and the role of innovation in advancing productivity. The book is intended to raise awareness regarding the importance of entrepreneurship and innovation in rebuilding Egypt's economy on strong foundations with potential for further development and faster growth.

We do not claim to have put together a compilation of all the conference proceedings, nor have we undertaken a comprehensive coverage of all issues pertaining to entrepreneurship and innovation in Egypt. Rather, we have included a series of selected insights that collectively advance knowledge on entrepreneurship and innovation in Egypt. We have placed special focus on embedding creativity and innovation in the educational system, addressing the challenges facing the entrepreneurial ecosystem, enhancing social entrepreneurship, providing linkages between academic research and applied/industry needs, and highlighting the relation between innovation and economic growth.

The book is composed of nine chapters written by scholars and practitioners in the areas of entrepreneurship, innovation, and development. In chapter 1, "Entrepreneurs as Heroes of Development," Zuhayr Mikdashi explores the quintessence of entrepreneurship. He first identifies the components of an "entrepreneurial hero," defined as the innovative entrepreneur with the ambition of changing things for the better. Mikdashi then examines entrepreneurs' motives in running their businesses and the relation between these motives and business performance.

In their chapter "Facilitating Entrepreneurship as a Catalyst for Change," Shailendra Vyakarnam and Shima Barakat evaluate recent attempts at building graduate-level entrepreneurship education programs, with a particular focus on the University of Cambridge, in the United Kingdom. The authors look into ideas and applications of entrepreneurship programs over the past fifteen years and their potential for promoting entrepreneurship as a catalyst for change in the Arab world.

In chapter 3, titled "Putting the Horse before the Cart: Understanding Creativity and Enterprising Behaviors," Andrew and Kathryn Penaluna challenge the premise that business skills and approaches should lead enterprise education. The authors emphasize that innovation and creativity are at the heart of the entrepreneurial process, and that these aspects need to be better understood.

In chapter 4, "Entrepreneurial Universities in Egypt: Opportunities and Challenges," David A. Kirby and Nagwa Ibrahim move to define what is meant academically by "an entrepreneurial university." Based on fieldwork research, they assess how Egyptian universities perceive the concept of an entrepreneurial university, how private and state universities implement entrepreneurship education, and how both can be extended and improved in the future.

In the fifth chapter, "Varieties of Entrepreneurs: The Entrepreneurship Landscape in Egypt," Ayman Ismail and Sherif Yehia review the entrepreneurial landscape in Egypt as one of the key enablers of economic development, job creation, and poverty alleviation. They cover various types of entrepreneurship, including high-growth, innovation-driven enterprises, micro, small, and medium enterprises, and social enterprises. The authors identify challenges facing entrepreneurs in key areas including access to finance, education, and training; support systems; research and development (R&D) transfer; and the regulatory framework.

In chapter 6, "Entrepreneurs in the 'Missing Middle': Know your Funding Options," Adel Boseli identifies the funding options available to Egyptian entrepreneurs and aims to help them select options that are most suitable for their businesses. This involves choosing from the pool of friends and family, business plan competitions, incubators, angel investors, crowdfunding, venture capital, and private equity.

In chapter 7, "Building a University-centered Entrepreneurship Ecosystem: A Case Study of the Entrepreneurship and Innovation Program at the American University in Cairo," Ayman Ismail presents a case study of a university initiative to contribute to building an entrepreneurial ecosystem in an emerging economy. Ismail provides a review of activities of the Entrepreneurship and Innovation Program at AUC in six key areas: entrepreneurs, ideas, networks, mentors, funding, and startups.

The chapter identifies the best practices and lessons learned in university entrepreneurship programs.

In chapter 8, "Schumpeterian Entrepreneurs, Total Factor Productivity, and Institutions: Firm-level Data Analysis from Egypt," Karim Badr explains what the term 'entrepreneur' means, looking at its Schumpeterian, neoclassical contexts, before analyzing firm-level data from Egypt that tests the relationship between innovative firms and productivity. He looks into the determinants of innovation in Egyptian firms. He stresses the importance of competition, research and development, education, governance, and access to finance for both productivity and innovation.

The book ends with a personal account from Khaled Ismail, an entrepreneur and academic. Ismail provides an innovative analysis of Egypt's January 25 revolution viewed through an entrepreneurial lens. The journey from the desire to solve a problem to incubation, then market acceptance, the struggles over ownership and decision-making structures followed by the lessons learned from failures. He also discusses the environment for startups and young entrepreneurs in Egypt over the past three years.

We hope that this compilation contributes to the conversation on issues relating to entrepreneurship and innovation in Egypt. Our authors have proposed some answers but they have also raised questions. For example, what drivers are needed to nurture and sustain the entrepreneurship ecosystem in the country? What policy changes are needed for universities to promote economic and social development in Egypt? In what ways can entrepreneurial educators foster creative thinking and innovation? How can universities in Egypt foster education *for* entrepreneurship, as opposed to merely education *about* entrepreneurship? We expect these and other pertinent issues to be the focus of further research and deliberations in Egypt at this juncture of the country's development.

Notes

1   International Monetary Fund, *Arab Countries in Transition: An Update on Economic Outlook and Key Challenges* (Washington, DC: International Monetary Fund, 2014), http://www.imf.org/external/np/pp/eng/2014/040914.pdf; World Bank, *MENA Quarterly Economic Brief* (Washington, DC: World Bank, 2014), http://www.worldbank.org/content/dam/Worldbank/document/MNA/QEBissue2January2014FINAL.pdf.
2   World Bank, *MENA Quarterly Economic Brief.*
3   World Bank, *Egypt Overview* (Washington, DC: World Bank, 2014), http://www.worldbank.org/en/country/egypt/overview.

4   Sara Aggour, "Over 50% of Egyptian Youth Are Poor: CAPMAS,"
      Daily News Egypt, August 12, 2014, http://www.dailynewsegypt.
      com/2014/08/12/50–egyptian-youth-poor-capmas/

1. Entrepreneurs as Heroes of Development

Zuhayr Mikdashi

Prologue[1]

Business creators and developers come from a wide spectrum of backgrounds. They belong to different age groups, social classes, ethnic communities, cultures, nationalities, and genders.[2] Some analysts use a broad definition of entrepreneurs that includes all people who create, own, and manage new businesses—whether their business is innovative or imitative (that is, copying existing enterprises, like groceries, hairdressers, coffee shops, pubs, fast food, or corner shops). Other analysts use a selective definition that limits entrepreneurship to innovative or 'Promethean' entrepreneurs who have an ambition to change things for the better. The selective definition is used in the following analysis as an *ideal benchmark*. It confers a heroic status on entrepreneurs for their achievements. Other profiles of entrepreneurs discussed in this chapter are to be evaluated by reference to the said benchmark.

Joseph Schumpeter's seminal works on entrepreneurship are an often-cited reference point in many interpretations of the entrepreneur's principal function. He considered entrepreneurs to be creators–innovators whose impact is revealed in how new products or processes replace existing ones. He argues that

the function of the entrepreneur is to reform or revolutionize the pattern of production . . . by exploiting an invention or, more

7

generally, an untried technology possibly for producing a new commodity or producing an old one in a new way; by opening up a new source of supply of materials or a new outlet for products, by reorganizing an industry and so on.[3]

It is important to note that new products and processes can widen the spectrum of choices available to users or consumers without necessarily destroying the demand for existing products. For example, artisan works produced using traditional methods can remain in demand along with similar wares produced using modern technologies, with groups of consumers interested in buying either or both. In exploring the quintessence of entrepreneurship, part 1 of this chapter attempts to identify the main characteristics of an entrepreneurial hero and proposes a composite benchmark. Part 2 examines the primary motives of entrepreneurs in the governance of their businesses, and the relevance of these motives to the overall business performance. Part 3 offers a case for illustrative purposes showing how the entrepreneurial hero can contribute to the well-being of his community and its overall economic performance. Finally, part 4 examines the potential role for academia and regulatory systems in the development and facilitation of entrepreneurship.

Part 1: A Composite Benchmark for the Entrepreneurial Hero

Entrepreneurs can be evaluated by multiple criteria. This chapter proposes a few fundamental and interdependent clusters of characteristics as being pivotal for constructing a benchmark for the entrepreneurial hero. That benchmark consists of five interconnected, mutually reinforcing pillars that could serve to rank entrepreneurs.

The first pillar of an entrepreneurial hero is a cluster of personal characteristics reflecting an insatiable curiosity combined with a flair and intuition for spotting or creating opportunities. The entrepreneur does not passively wait for opportunities, but actively seeks them out. Thinking "outside the box," heroic entrepreneurs have an instinct and talent for disregarding conventional wisdom. Often described as unorthodox, entrepreneurs have the courage to overcome prevailing complacency by advocating bold ideas. They are capable of extracting profit-making solutions out of current problems. Entrepreneurs can use novel technologies or ameliorate existing ones. An entrepreneur need not be an inventor, but he or she has the capacity to understand current technologies and put them to optimal use. In seeking higher levels of efficiency and performance, an entrepreneur's inventiveness can come from the reorganization or restructuring of his or her business activities.

The second pillar of an entrepreneurial hero is his or her sound vision and strategies. He/she could start with good rules of thumb, followed by an evaluation of a project's strengths, weaknesses, opportunities, and threats (SWOT), then progress to a thorough, detailed quantitative and qualitative analysis to confirm or invalidate an initial instinctive appreciation. A sound multi-year business plan extending over three to ten years depends on the nature of the project and the expected development period. The business plan presents the objectives, priorities, and timing for the implementation of the project. Carefully estimated information is needed, notably regarding investments in human and material resources, organization and strategy, expected disbursement of funds and generation of revenues, and other key business data. Continuous evaluation of activities should serve to identify and remedy gaps, shortcomings, and errors; make adjustments in the business plan to overcome hurdles as they arise; improve efficiency and performance; and address changes in competitive conditions and the sociopolitical environment.

The third pillar of an entrepreneurial hero is intrepidness and the refusal to procrastinate. He/she is tenacious and enduring in his/her unwavering commitment to the project's goals in the medium and long term, regardless of the vagaries of unfolding circumstances. Intrepidness is fueled by a zeal to realize his/her goals, even at the cost of material resources and personal rest and leisure time. He/she is admired for accomplishing his/her projects even when surrounding uncertainties or obstacles seem foreboding. Perseverance need not preclude flexibility when justified, however.

The fourth pillar of an entrepreneurial hero is his/her capacity to translate an embryonic idea into an enduring business success. Starting with a blueprint, the entrepreneur's next stage is to create a workable prototype. Subsequently, he/she would have to efficiently produce the refined trial model for profitable commercialization. When proposing a concrete solution, the entrepreneur relies on his/her insights and inventiveness. Honed by common sense, careful reasoning, and relevant experiences, his/her performance will be measured by sales growth and long-term revenues.

The fifth pillar of an entrepreneurial hero is his or her ethically based conduct vis-à-vis all stakeholders. Staunchly observed ethical principles are a vital, strategic asset that enhance the economic performance of the enterprise and boost image and goodwill among customers, employees, partners, and regulatory agencies. Honesty, especially at the leadership level, is a major component of that vital asset.[4] The heroic entrepreneur respects ethical principles in supplying products, in the treatment of workers–collaborators, in the balanced allocation of benefits and costs among stakeholders, and in abiding by legal and civic responsibilities. He/she shuns predatory practices, such as deriving wealth or profits by exploiting users, consumers, workers, or subcontractors, or by harming the ecosystem. By complying with good governance and honoring commitments, an entrepreneur builds stakeholders' trust and motivates them to emulate him in furthering the prosperity of the enterprise. After having looked at the five pillars that make an entrepreneurial hero, we will go on to explore the three different types of entrepreneurs and how they are related to socioeconomic performance.

Part 2: The Three Profiles of Entrepreneurs

We explore hereafter three main categories of entrepreneurial profiles. Each of the different entrepreneurial profiles – that is to say, the motivation behind entrepreneurial activity – influences the socioeconomic performance of the entrepreneur, which makes it important to identify the factors that describe each. They include the egocentric entrepreneur, who is prompted essentially by self-centered personal control and gain; the community-driven entrepreneur, geared toward serving one or more selected communities; and the stakeholder-inclusive entrepreneur, motivated by the balanced split of responsibilities, costs, and benefits.

Self-centeredentrepreneurs tend to have the dual objectives of maintaining full control over their business and maximizing their own benefits. They could be considered particularly individualistic and materialistic, and prone to the insatiable pursuit of wealth for its own sake. With a proclivity for elitist self-esteem that occasionally verges on narcissism, these entrepreneurs focus primarily on their own interests.

Many researchers have acknowledged that the pursuit of personal gain is a high-intensity motivation for business creation and development, and can produce outstanding performance. The fundamental challenge, however, is to guard against the possibility that this pursuit could lead to the slippery slope of hubris and greed as business leaders become mesmerized by the possibility of gaining more than they contribute to the value pie. This in turn can lead to impulsive, unjustified risks, short-term thinking that favors proximate benefits at the expense of more significant long-term benefits, and/or predation by encroaching on stakeholders' legitimate rights. Egocentric entrepreneurs could encounter difficulties in recruiting and retaining talented collaborators, forging stable relationships with reliable partners, and winning the loyalty of consumers and users, as these parties prove reluctant to fall prey to the entrepreneur's ambition.

Community-driven entrepreneurs generally espouse the goal of service to a targeted community. Also called social entrepreneurs, they are distinguished from unpaid volunteers or charity providers in that they expect to be reasonably remunerated for their services—though usually at more moderate levels than self-centered entrepreneurs, who are bent on their own profit maximization. The community-driven entrepreneur's main objective is to generate benefits for the targeted community in the form of profitable business activities, jobs, and needed infrastructure.

Several scholars have analyzed the motives, strategies, and accomplishments of entrepreneurs who are driven by the spirit of benefiting communities. Community-driven entrepreneurs generally lead frugal lifestyles as compared to their self-centered peers. Some derive personal satisfaction from their activities, even if just the pleasure of being recognized for their actions, while others do not seek publicity and are genuinely low-key. Heightened recognition and prestige for these entrepreneurs' valuable role could serve to spread the spirit of emulation.[5] One should recognize, though, that business activities favoring a specific community are not necessarily equivalent to serving the common good, in cases where other, more needy groups are excluded from these activities.

Stakeholder-inclusiveentrepreneurs are recognized for their consideration of all their stakeholders. They are typically driven by a culture of inclusiveness, fairness, and equal participation. In so doing, these entrepreneurs attract and keep talent, earn the loyalty of their collaborators, build reliable business relations with partners and contractors, and satisfy consumer expectations.

Modern organizational strategic management theory emphasizes the positive role of balanced participation among stakeholders:[6]

> The conventional notion that the corporation should create wealth only for its shareowners is incorrect. The corporation should be redefined to emphasize its relationships with and responsibilities toward *all* stakeholders, both voluntary and involuntary . . . . The stakeholder model of the corporation fits in with broadly accepted normative and ethical considerations.[7]

Stakeholder-inclusive entrepreneurs are characterized by a sense of collective responsibility beyond their own narrow self-interest. They are concerned with the proportionality of benefits distributed among their stakeholders. Although difficult to calibrate, attain, and/or maintain durably in a dynamic world, a managerial philosophy based on the entrepreneur's evenly sharing the value pie with stakeholders strengthens the zeal of all parties. It also reduces conflict between management and workers, enhances productivity, eliminates despondency or neglect, and protects against stress, anxiety, and sickness. Empirical evidence has shown that business leaders who embrace process fairness reduce employee discontent, enhance efficiency, generate support for new strategic initiatives, and promote an environment of creativity and performance. The balanced and fair distribution of the net value among stakeholders—as commensurate with each party's contribution over time—is nevertheless a difficult task.

The stakeholder-driven entrepreneur's challenge is to marshal the energies of all potential contributors toward helping the enterprise accumulate wealth by realizing synergies among stakeholders. This will be achieved if all concerned parties feel they are fairly treated by the entrepreneur at the helm of the decision-making process with respect to transparency, decision-making, distribution of net value, and other critical matters. Some empirical researchers conclude that stakeholders are more likely to contribute their best efforts to value creation when they anticipate that they will receive their fair share during the process of value distribution.[8]

The following list presents the major stakeholders in a business entity—here an entrepreneurial venture—followed by their principal concerns:

1. Entrepreneurs and partners: raising the net value of their venture
2. Inventors and innovators: proper rewards for their talents
3. Customers: quality/price mix, service, security of supplies
4. Managers: compensation packages
5. Employees and workers: adequate pay, working conditions, tenure, and training
6. Creditors: safety of principal and interest
7. Suppliers and subcontractors: mutually beneficial relations
8. Communities and authorities: jobs, revenues, governance, and sustainable prosperity

The above list could be extended to include other stakeholders, such as non-governmental organizations (NGOs) and the media, whose mission is to monitor, analyze, report, and evaluate the conduct and performance of entities under their purview. By exercising their activities in an independent, objective, and impartial spirit, NGOs and media can be considered legitimate stakeholders in their role as watchdogs. They have to be distinguished from partisan or self-interested lobbyists.[9]

Distinctions between the businesses started and managed by these three different entrepreneurial types are not always clear-cut. Indeed, some entrepreneurs may change their behaviors over time, and some could have hybrid or ambivalent motives. Each of the above-mentioned behavioral styles can be considered rational, or can be rationalized by the entrepreneurs concerned.

The following section, using the example of a village in Senegal, seeks to highlight how the above-mentioned types of community-driven and stakeholder-driven entrepreneurship can work in praxis, putting the entrepreneurial spirit to work for the betterment of the community and overall socioeconomic performance.

Part 3: Case Study: Overcoming Distress
Background
Much of the world's population lives in economically, socially, and/or politically desperate conditions. Economic destitution can be ascribed to a nation's diminished economic assets and the consequent lack of opportunities. National assets generally cover three categories: built or manufactured physical capital, such as machinery, buildings, and infrastructure; human capital, such as education and skills; and natural resources, including arable land, forests, clean water and air, fuels, minerals, and a healthy environment. Some of these assets, such as mineral resources, are depletable, while others, such as clean water and air, cannot readily be priced. Shadow prices have to be established for those products whose market prices do not integrate their negative externalities, such as pollution, or their positive externalities, such as hydroelectric dams offering benefits like flood control and irrigation.[10] I propose a fourth category for an all-inclusive wealth of nations, namely, the observance of fundamental human rights principles.

Some countries have natural wealth, as defined above, but their populations live in fear for their security and livelihood, as they are subject to the tyranny of rulers who eviscerate their national economies. Many are forced to leave their homes in search of jobs and security, risking their lives in the hope of a safer haven. Natural disasters, civil wars, industrial-economic disasters, and abuses of power force them to seek sanctuary from persecution, violence, and deprivation. Lured by the mirage of hospitable developed societies, thousands of migrants suffering from endemic unemployment have been desperate to reach Europe, North America, Australia, and other developed regions.

Should they manage to survive the crossing of forbidding territories, many perish in harsh physical or social environments, and are exposed to dire exploitation and abuse in countries of passage or destination.

The hardships that impel populations to flee their home countries do not only affect unskilled laborers. They also impact skilled and highly-educated populations. In the 'brain drain' phenomenon, many young, well-educated people move from their homelands to developed countries, where opportunities for work and creation are much more attractive.[11]

At the same time, though, numerous individuals seek to take charge of their lives by setting up their own businesses. Policy makers, business leaders, and academics have praised the significant role of the entrepreneurial self-employed in the category of micro to small businesses, as they generate benefits for themselves and for stakeholders through innovation, job creation, GDP growth, and the spread of wealth.[12]

The following case study illustrates the role of entrepreneurs in trying situations. It highlights the positive impact that community-driven as well as stakeholder-driven entrepreneurship can have on its environment.

Creators of a community hub in Senegal

Moving against the general trend of migratory flow from poor to richer societies, the following case illustrates a 'brain gain' in a destitute region. It is the story of Serigne Babacar Mbow and his wife Pascale Blanchard, whose achievements are due to creative ideas, ethical values, and unwavering commitment, with the support of a dedicated circle of like minded people. During his youth in the 1960s, Babacar fought for independence and social reforms in his home country of Senegal. He was imprisoned several times with his comrades by the French colonial authorities, before leaving to study in France.

The Mbow couple was living comfortably in Paris, one of the world's most attractive metropolises, until the early 1980s, when Babacar was spurred by a vision to transform the native village of his forefathers, Ndem (located in Senegal's southeast bush land, 120 kilometers from the capital, Dakar), from a drought-stricken deserted locality into an economically viable community. Babacar is a descendant of the founder of Ndem. His devoutness has led him to offer spiritual guidance for the Ndem community, within the mystical Mouride brotherhood that extols industriousness and lofty values of self-help, solidarity, tolerance, peace, and respect for others. The brotherhood preaches the sanctification of toil and discipline, considered to be integral to piety insofar as they protect humans against the indignity of begging.[13] Following Babacar's vision, Babacar and Pascale had to address some fundamental questions with crucial implications for themselves and their potential collaborators, chief among them:

1. Should they carry on with their life of relative comfort in Paris, or should they rise to the challenge of the noble vision of rejuvenating a community in dire poverty?
2. Could the bicultural couple (they had one child at the time) integrate in the local community?
3. Would they be able to bear a life of toil and moderation for long in their intended new home, with its harsh, arid physical environment? Would a decent livelihood be possible for them, as well as for those who would join them in due course?
4. Would Pascale's family support her decision to leave France?

They made their remarkable decision to leave Paris for Senegal during a period when thousands of African people were fleeing for the north (notably western Europe), lured by the possibility of decent work and a better quality of life. The couple's bold move south was prompted by their rebellion against the fate of poverty and the many indignities it brings in its wake, and the belief that a decent living can be honestly obtained through entrepreneurship, innovation, and hard work. The Parisienne Pascale probably had to make a bigger leap than Babacar when envisaging a fulfilling life in an African community bereft of its young people, with most families fleeing from misery. But Pascale enthusiastically accepted the mores and traditions of her adoptive community, and changed her name to Sokhna Aïcha Cissé. Her courage and stamina were exemplified by a life devoted to her adopted community in the harsh climate of the Sahel region. The Mbow couple started their new life in Senegal by constructing a hut for their family (which eventually counted six children) on a plot of bush land offered by the welcoming villagers.

To win back those who had immigrated to Dakar or to foreign lands, Serigne Babacar Mbow, supported by his wife and his followers, sought to reinvent Ndem as an economic hub. They started a profitable African clothing business. For their venture, Babacar's wife provided her know-how and her capital, starting with the sewing machine she brought from France. Marketing support from Pascale's mother through a boutique outlet in the Marais, a neighborhood in the heart of Paris, was invaluable. The boutique sold articles and wares produced by the couple and their fellow villagers.[14] The Mbow venture has been emulated by other villagers, thereby building Ndem's reputation for quality artisanal hand-made cotton clothes and household wares for sale in Africa and in boutiques in Europe and North America.

Babacar brought fourteen villages together by founding the Association of Ndem Villages (ANV) in 1985; he became its president, while his wife served as associate president. In 2006, with 4,600 members, ANV changed its status from that of a cooperative to that of an NGO to promote collective investments for the benefit of all community members. A portion of the artisans' profits was collected to build the community's infrastructure. These investments included a health center, a school compound, deep wells in aquifers for potable water, microfinance, and solar and biomass energy.

The villagers initially focused on weaving and cloth-making, primarily of multicolored patched garments inspired by African traditions and with designs from local artists. Their activities expanded to cover a variety of crafts and trades, such as embroidery, leather works, furniture, and other housewares. They used African cotton, natural dyeing, and other ecological materials from the region. Ndem's activities expanded even further over the years to satisfy local needs for vegetables and cereals produced with techniques suitable for an arid climate, thereby increasing the local community's economic autonomy and wealth. Ndem thus evolved from a deserted township to an oasis of creativity thanks to the leadership of the Mbow couple, whose values and strategies won over villagers bent on vanquishing destitution and halting the painful exodus of their people. The Mbow couple's vision saved Ndem from being a ghost town. Many other once-thriving communities were devastated by the loss of key resources due to deforestation, depleted water supplies, exhausted commercial mineral resources, the obsolescence of products, the encroachment of the desert on arable land, and other natural or man-made disasters.

The Mbow couple's success owes much to their unity of purpose, guided by love for lofty ideals of human rights and honest toil, and refusing to let their community succumb to the humiliation of poverty and dependence on charitable handouts. Other significant factors that contributed to their success were their active, unrelenting personal commitment to the venture they launched, from conception to full realization; their genuine concern for the equitable treatment of the wide circle of stakeholders; and the reliance on competitiveness derived from sustainable local inputs and skills, along with artistic production that appealed to diverse tastes in various markets. This example clearly highlights how the entrepreneurial spirit of one couple turned around the fate of an entire area, exemplifying the importance of fostering community-driven as well as stakeholder-driven entrepreneurship. Thus, the next section addresses the question as to how academic institutions and regulations by public authorities can foster this kind of entrepreneurship.

Part 4: Enhancing and Spreading Entrepreneurial Capacities

Academic institutions and public authorities have an important role in nurturing entrepreneurial efforts. Institutions of learning and training can scrutinize and evaluate business models to increase efficiency and heighten attunement to changes in society and diverse local conditions. Public authorities can facilitate entrepreneurial activities through various means, such as assisting in literacy and skills campaigns, building basic infrastructure, easing access to funding, streamlining procedures, and empowering citizens to manage local affairs.

Role of knowledge and training centers

In recent decades, the world economy has witnessed an upward trend in knowledge-intensive business activities. Education, research, experience, and competencies all represent the stock of human capital available in any economy. The higher the quality and the size of that human capital, and the more efficiently it is deployed, the higher the level of innovation and business opportunities. Learning hubs are notable for favoring innovation, economic growth, and societal progress.[15]

Leadership in the twenty-first century's business sector calls for interdisciplinary approaches that can offer decision makers a competitive advantage in opening up new vistas, introducing novel ideas, and applying efficient tools to complex challenges. Increasingly, universities offer interdisciplinary diplomas in areas such as business and public administration, natural sciences, and technologies, which are suited for managerial positions in firms dependent on scientific knowledge for their sustained growth.

Possible academic industry initiatives that can help entrepreneurial talents flourish include practical learning within dynamic enterprises, and industrial science parks and innovation clusters hosted by higher centers of learning. With respect to practical learning, programs typically have the following characteristics:

1. Students from business, engineering, sciences, and other disciplines spend a few weeks working with an entrepreneurial business, offering help within their field of study, as requested by the entrepreneur.
2. They operate under the dual guidance and supervision of their professor and the host entrepreneur or an associate.
3. By the end of their sojourn in the enterprise, they prepare a report on their activities (certain confidential details can be omitted or made available solely to a designated direct supervisor, to protect proprietary information).
4. They receive academic credit for productive time with the enterprise, and are graded by the faculty member in consultation with the entrepreneur or his/her associate.
5. Host enterprises are identified by individual students, faculty, student associations, professional groups, and others.
6. Students do not expect any form of monetary remuneration.
7. Possible contributions by host enterprises help defray the costs of the practical learning program, covering necessities like transport and meals.
8. The possibility of matching two or more students from different disciplines is likely to be productive, and deserves consideration.

Several academic institutions have also developed industrial parks or startup incubators for graduates who aspire to become entrepreneurs. These usually offer meeting spaces, laboratories, and computing centers that are made available at reasonable terms to alumni who are creating and developing their businesses. Graduates not only benefit from the available technical and office facilities but can also call on the advice of faculty members and researchers. Should these startup companies take off, the academic institution is likely to derive some positive benefits. Owners of a startup company typically leave the incubator park after an agreed period to allow for a fresh group of aspiring entrepreneurs to move in.

Industrial parks have demonstrated usefulness for students and alumni who show an entrepreneurial spirit and hope to commercialize their innovations. These up-and-coming entrepreneurs benefit from academic contacts in the pool of talent and expertise available in the industrial park; the tools put at their disposal (notably, direct access to infrastructure such as laboratories and computing centers); and privileged contacts with outsiders (investors, lawyers, accountants, and other relevant professionals). As they are hosted at these parks, these grassroots entrepreneurs are generally not burdened by excessive fees or other inroads into their financial gains. Knowledge, in all its varied dimensions, should lead to cross-fertilization, thereby broadening and deepening benefits for all parties concerned.

Certain features of these industrial parks merit careful consideration in order to address challenges and to accommodate particular situations:

1. When setting up incubators for startups, higher centers of learning in developing economies would be well advised to pool their resources for greater efficiency in building well-equipped industrial parks. By sharing efforts, resources, and costs, academic institutions can create better-equipped facilities for potential entrepreneurs, ultimately facilitating the creation of new entrepreneurial entities. It could also prove productive to focus on a cluster of related innovations in a few select industrial sectors when choosing which startups to host, in order to stimulate combined technological and business excellence. Innovative entities with high-growth potential should be privileged. A jury panel comprising seasoned entrepreneurs and academics could separate the wheat from the chaff among the numerous candidates seeking to join the industrial park as potential entrepreneurs. Candidates should provide carefully prepared and solidly documented multi-year business plans—possibly extending up to ten years for certain proposed projects—for the jury panel to evaluate their relative attractiveness. To give substance to innovative entrepreneurship, industrial parks need to facilitate access to venture capital or private equity funding. Investors should be invited to contribute seed capital to startup ventures of their choice. Not all countries have these funding sources, however. Moreover, certain entrepreneurs shun third-party investors, since they want to keep exclusive control over their projects.[16] A university-sponsored investment vehicle could consider a minority interest in selected projects, with the expectation of deriving benefits from their sale in later years. Alternatively, the vehicle could act as a facilitator that attempts to attract new investors.[17] Industrial parks should set deadlines for entrepreneurs to develop their projects from the early-stage blue print to a prototype of a sustainable, viable project. Flexibility may be called for, though, depending on the nature of individual projects.

Higher centers of learning and research are arguably best equipped to provide current and potential entrepreneurs with the appropriate know-how, and familiarize them with the nature, utility, and limitations of various managerial methods or tools relevant to their enterprises. One could also argue that academia is well placed to offer exposure to specific business issues—such as risk, ethics, and good governance—through workshops, seminars, conferences, peer exchanges, and other channels. Academia can thereby impress on business participants the advantages of abiding by the best practices.[18]

Regulatory incentives

The state has an important role in stimulating the economy by encouraging its entrepreneurs. The latter generally start small, a situation which poses significant challenges in periods of economic slump. In these periods, firms are wary of hiring, lest they get stuck with employees for whom they have no work and whom they cannot readily lay off. Small firms in particular have relatively limited financial resources to retain well-qualified workers during the gap between a slump and a recovery.

To promote desirable objectives, such as higher levels of employment, growth, and well-being, policy makers can offer various incentives to entrepreneurial individuals who create businesses. Most governments in advanced or emerging market economies have created special facilities or vehicles that aim to propp up micro, small, and medium-sized enterprises (MSME) through diverse technical, financial, and material resources, and/or by alleviating administrative and fiscal burdens.[19] Policy makers, business leaders, academics, and others have praised entrepreneurs for taking control of their lives by setting up their own businesses. They particularly welcome entrepreneurs who generate benefits for their stakeholders through innovation, job creation, GDP growth, and the spreading of wealth.[20]

Small-sized enterprises face disadvantages in terms of diseconomies of scale—or higher unit costs in view of relatively small, sub-optimal output—and the small business owner–manager's inability to singlehandedly focus on all the functions of his/her business. He/she needs to be relieved from certain tasks by recruiting subordinates with specialized skills in areas such as product quality, marketing, administration and finance, technology and logistics, accounting and controls, and human resources. Given a dearth of financial resources to recruit the requisite talent, the owner–manager of a small enterprise often handles several of these activities.

Other interrelated challenges encountered by the small enterprise are a lack of product and geographical diversification, and dependence on very few customers. Funding is constrained in view of a relatively low equity cushion to absorb possible losses or to invest in expansion. Inadequate reporting procedures on functions, investments, or lines of business hamper analysis for assessment, accountability, and controls. Stringent rules for the dismissal of redundant workers, such as large severance packages borne solely by small enterprises, are onerous. Any of these disadvantages can prove to be the Achilles heel of a small company.[21]

Regulatory systems contribute to entrepreneurial dynamism if they encourage personal initiative and risk-taking. Entrepreneurs can be thwarted, however, by various socio-administrative hurdles that hamstring their creativity. They need to be stimulated by business-friendly environments with an appropriate regulatory framework that avoids bureaucratic red tape, and provides stability and predictable measures. For example, a regulatory framework that encourages entrepreneurial individuals to try new initiatives, despite previous non-productive starts, bolsters innovation and value creation in the long term. Insolvency regimes should not treat honest, insolvent entrepreneurs like fraudsters. Barring criminal charges, showing forbearance toward entrepreneurial individuals' past failures could open up possibilities for them to learn from past errors, bounce back, and succeed in creating new, performing businesses.[22] Legislation providing small business borrowers with bankruptcy protection through insurance or guarantees, for example, encourages risk-taking and promotes an entrepreneurial culture.[23] The World Bank Group has created an indicator of ease of doing business that includes the measures necessary to start a new business, and the time and costs necessary to register and launch such a business.[24]

Firms whose activities are declining should have the flexibility to prune their personnel, with adequate notice and payoff linked to salary and service for redundant employees. Stringent job protection regulations might deter budding entrepreneurs from recruiting, lest they become saddled with wage bills they cannot reduce in periods of economic difficulty.[25] Safety nets provided by public authorities or industry can help those who lose their jobs. In periods of economic restructuring due to a decline in business activity, employees should ideally be offered retraining—if they have the capacity—to transition from dying jobs into newer ones. They could also be assisted in moving to regions with higher employment opportunities. To determine which entrepreneurial enterprises should be incentivized, one needs first to identify the categories of enterprises that produce superior contributions to society. This has to be followed by the examination and implementation of appropriate measures to promote this category of enterprises. Among the pioneering studies in this field, Global Entrepreneurship Monitor (GEM) reports have identified "high-expectation entrepreneurial firms" that are less than four years old and expect to achieve rapid growth in employment size—that is, to employ at least twenty employees within five years' time. These firms are responsible for 80 percent of newly-created jobs, according to a survey of more than forty-four countries worldwide.[26] One also needs to take into consideration the net job creation of startups, after deducting the estimated job destruction caused by the same startups.[27]

For governments vitally concerned with high-performance objectives of reducing unemployment, promoting innovation, enhancing added value, and boosting exports or growth, selective support of policy measures is needed. These stimuli include easing access to credit at reasonable terms, reducing administrative procedures, and temporary tax breaks or subsidies in favor of the above-mentioned "high-expectation" entrepreneurial enterprises. This is the policy position of many governments. The French government, for example, supports the most dynamic enterprises—meaning those whose annual growth in turnover exceeds 34 percent—among independently-owned or -controlled small to medium-sized enterprises (SMEs). These enterprises employ between five and 250 salaried employees, with a turnover volume of less than EUR50 million. These companies are identified with the "gazelle" label of distinction, and benefit from the seasoned advice of experienced business leaders and financiers.[28] Moreover, the French governmental agency OSEO provides financial support in the form of direct funding or guarantees with attractive terms to innovative SMEs.[29]

Independent nongovernmental and nonprofit organizations also can help entrepreneurial projects with a considerable growth potential. This is the case with the US-based NGO Startup America Partnership, which provides budding entrepreneurial companies with valuable resources—including advice and mentorship, access to critical services at reduced costs, assistance in recruiting, training, and retaining talent, acquiring new customers, introduction to thought leaders and major corporations as potential partners or customers, and information on appropriate sources of funding. Startup America has an extensive network of grassroots and seasoned entrepreneurs, universities, independent local talents, established large companies, and governmental agencies.[30] The propagation of high-performance entrepreneurs should contribute to society's well-being.

To maximize their creative potential, entrepreneurs need to operate in free, but soundly regulated, open markets. Such business-friendly markets respect the principles of private management and state governance. To perform their wealth-creating activities, entrepreneurs cannot be shackled by bureaucracy that delays or blocks their actions. Understandably, the state has to perform its functions of protecting users and consumers from monopolistic practices, controlling the quality of products, inspecting and sanctioning fraud, assuring the health and safety of working conditions, regulating pollution, and other public-interest functions.

Epilogue

Entrepreneurs thrive in a society that encourages personal initiative and risk-taking and shows forbearance toward initial business setbacks, with the expectation that individuals are able to bounce back and succeed after learning from past mistakes. A liberal regulatory framework that encourages entrepreneurial individuals to continue innovating despite past failures bolsters their value creation potential over the long term. Adequate reservoirs of both managerial and technical expertise and accessible financing at reasonable terms facilitate entrepreneurial endeavors.

Tax privileges and access to investors are not the only incentives to boost entrepreneurship. Supportive structural conditions are also needed. These include unencumbered, stable legal frameworks for the establishment, operation, or exit of businesses; an impartial judiciary system; and efficient financial markets. Regarding the latter, it is important to note the advantages of stock markets that allow innovative companies to realize value creation through initial public offering (IPO) exits, in addition to other venues, such as their sale to existing industrial companies or to financial groups.

Countries that foster investment in human capital can improve their economies and change from producing low value-added commodities to higher value-added ones. This requires several preconditions, such as encouraging students and researchers to pursue scientific and technological fields that could breed inventions and innovations; adequate systems of personal motivation, like appropriate working conditions and compensation incentives; the absence of constraints or burdens that shackle original, innovative projects and novel entrepreneurial endeavors; and a propitious sociopolitical environment characterized by peace and stability with open, non-discriminatory access to business opportunities.

An increased stock of know-how and experience is likely to contribute to entrepreneurs' success in enhancing the value of their enterprises.

Education can stimulate predisposed individuals to become entrepreneurs, providing them with appropriate tools for their actions. One should add that the macroenvironment in which potential entrepreneurs operate is crucial to enhancing motivation. Key positive factors in this environment include sociopolitical stability; sound economic, fiscal, and monetary policies; competitive markets; a fair judicial system; efficient regulatory agencies; and good governance.

The five-pillar entrepreneurial hero paradigm calls for metrics to permit policy recommendations. The performance of these policies depends on spotting and creating opportunities; sound vision and strategies; intrepidness, inventiveness, and performance; and ethical principles. Though lofty, these components can serve as the best guides. Since short-term personal gain is likely to remain a high-intensity fuel for entrepreneurial creativity and dynamism, vigilance is needed to guard against the possibility that the natural motivation of selfish gain and power could become excessive.

Notes

1   The author expresses sincere gratitude for insights he received during his participation in 2012 at the American University in Cairo's (AUC) Research Conference on Entrepreneurship and Innovation. This landmark event was organized and co-chaired by two eminent colleagues: Dr. Nagla Rizk, associate dean, Graduate Studies and Research, and director of the Access to Knowledge for Development Center at the School of Business; and Dr. Hassan M.E. Azzazy, leader of the Novel Diagnostics and Therapeutics Research Group at the School of Sciences and Engineering. Sincere gratitude is equally owed to Dr. Shawki Farag (chair of the department of accounting at AUC), who was an inspiring companion throughout the author's academic sojourn as distinguished visiting professor at AUC's Business School during the 2012 academic year. Dr. Farag has also offered valuable comments on an early version of this chapter. To these colleagues, along with their teams, I express my profound appreciation for their most valuable support and for their exemplary courtesies. Gratitude is also owed to Sylvia Zaky for her thorough editorial help.

2   For convenience, I use hereafter the masculine form for entrepreneurs to designate persons of either gender whose newly-created business could be individually or jointly owned and managed.

3   Joseph Schumpeter, *Capitalism, Socialism, and Democracy* (New York: Harper & Row, 1950), 132.

4   See Dan Ariely, *The Honest Truth about Dishonesty* (New York: HarperCollins, 2012), 255.

5   See Joel Brockner, "Why It's So Hard to be Fair," *Harvard Business Review*, March 1, 2006, 122–29; Leonardo Becchetti and Carlo Borzaga, eds., *The Economics of Social Responsibility: The World of Social Enterprises* (London:

Routledge, 2010), 258; Jost Hamschmidt and Michael Pirson, eds., *Case Studies in Social Entrepreneurship and Sustainability* (Sheffield, UK: Greenleaf Publishing, 2011), 452; Ehaab Abdou, *A Practitioner's Guide for Social Entrepreneurs in Egypt and the Arab Region* (Cairo: The American University in Cairo–John D. Gerhart Center for Philanthropy and Civic Engagement, 2010), 90.

6   See R. Edward Freeman, *Strategic Management: A Stakeholder Approach* (Boston: Pitman Publishing, 1984); R. Edward Freeman and John McVea, "A Stakeholder Approach to Strategic Management," Darden Business School Working Paper No. 01–02, 2001; Michael A. Hitt, R. Edward Freeman, and Jeffrey S. Harrison, eds., *The Blackwell Handbook of Strategic Management* (Blackwell Publishing, 2006).

7   James E. Post, Lee E. Preston, and Sybille Sachs, *Redefining the Corporation: Stakeholder Management and Organizational Wealth* (Stanford: Stanford University Press, 2002), 32–33, 376.

8   See Hossam Zeitoun, "Corporate Governance Modes, Stakeholder Relations, and Organizational Value Creation" (PhD diss., University of Zurich, 2011), 123.

9   See C.B. Bhattacharya, Sankar Sen, and Daniel Korschun, *Leveraging Corporate Responsibility: The Stakeholder Route to Maximizing Business and Social Value* (Cambridge: Cambridge University Press, 2011), 340; Dorothea Baur, *NGOs as Legitimate Partners of Corporations: A Political Conceptualization* (New York: Springer, 2011), 204; George F. DeMartino, *The Economist's Oath: On the Need for and Content of Professional Economic Ethics* (Oxford: Oxford University Press, 2010), 272.

10 See UNU-IHDP and UNEP, *Inclusive Wealth Report 2012: Measuring Progress toward Sustainability* (Cambridge: Cambridge University Press, 2012), 334.

11 See "Visas for Entrepreneurs: Where Creators Are Welcome," *The Economist*, June 9, 2012, 62, referring to Egyptian-born technology entrepreneur Mohamed Alborno, CEO of Crowdsway (www.crowdsway.com).

12 See Paul Vandenberg, "Poverty Reduction through Small Enterprises" (Working Paper, International Labour Office, Geneva, 2006), 59; IBM, *Global Innovation Outlook 2.0* (March 2006), 49; International Finance Corporation, "IFC SME Ventures Fund" (www.ifc.org).

13 See Tim Judah, "Senegal's Mourides: Islam's Mystical Entrepreneurs," *BBC News*, August 4, 2011.

14 Rose Skelton, "Senegal Clothes Maker Gives Faithful Jobs and Hope," *BBC News*, June 20, 2006, http://news.bbc.co.uk/2/hi/africa/5075534.stm.

15 See Soumitra Dutta, ed., *The Global Innovation Index 2012: Stronger Innovation Linkages for Global Growth* (INSEAD and World Intellectual Property Organization, 2012), 440.

16 See Zuhayr Mikdashi, *Progress-driven Entrepreneurs, Private Equity Finance and Regulatory Issues* (Basingstoke, UK: Palgrave Macmillan, 2010), 216.

17 Cf. Oxford Centre for Entrepreneurship and Innovation's Venture Club, Saïd Business School, www.sbs.ox.ac.uk.

18 See David Molian, "Entrepreneurial Value Creation: Are European Business Schools Playing Their Full Part?" (paper presented at the European Foundation for Management Development Conference, Maastricht, March 5–6, 2012), 16.

19 See European Commission, *Action Plan: The European Agenda for Entrepreneurship* (February 11, 2004), 20. Brussels, European Commission, *Minimizing Regulatory Burden for SMEs* (November 23, 2011), 14.

20 See OECD, *SMEs, Entrepreneurship and Innovation* (June 2010), 212.
21 See Adrian Beecroft, *Report on Employment Law*, United Kingdom Department for Innovation and Skills (October 24, 2011), 16.
22 See European Commission, *A Second Chance for Entrepreneurs: Prevention of Bankruptcy, Simplification of Bankruptcy Procedures and Support for a Fresh Start*, Final Report of the Expert Group (January 2011), 13.
23 See the publications of the intergovernmental think tank OECD on this subject. See also Jeremy Berkowitz and Michelle J. White, "Bankruptcy and Small Firms' Access to Credit," *RAND Journal of Economics* (Spring 2004): 69–84; Wei Fan and Michelle J. White, "Personal Bankruptcy and the Level of Entrepreneurial Activity," *Journal of Law and Economics* (October 2003): 543–67; "Life after Debt," *The Economist*, April 16, 2005, 72.
24 See World Bank, *Doing Business 2012*, www.doingbusiness.org
25 See "Companies and Productivity: Small Is Not Beautiful," *The Economist*, March 3, 2012, 12; "The Trouble with Small Firms," *The Economist*, March 3, 2012, 69; "Explaining Economic Weakness: The IMF v Beecroft," *The Economist*, May 26, 2012, 33.
26 See Erkko Autio, *2005 Report on High-expectation Entrepreneurship*, Global Entrepreneurship Monitor (2005), 52.
27 See Tim Kane, *The Importance of Startups in Job Creation and Job Destruction* (Kauffman Foundation of Entrepreneurship, July 2010), 8.
28 République Française—Direction du commerce, de l'artisanat, des services et des professions libérales, *Statut de la PME de Croissance—guide pratique* (Paris 2007), 38, http://www.bnf.fr/documents/biblio_tpe.pdf.
29 BPIFrance, "BPIFrance," www.oseo.fr.
30 See Startup America Partnership, http://www.s.co/#

Bibliography

Abdou, Ehaab. *A Practitioner's Guide for Social Entrepreneurs in Egypt and the Arab Region*. Cairo: The American University in Cairo–John D. Gerhart Center for Philanthropy and Civic Engagement, 2010.

Ariely, Dan. *The Honest Truth about Dishonesty*. New York: HarperCollins, 2012.

Autio, Erkko. *2005 Report on High-expectation Entrepreneurship*. Global Entrepreneurship Monitor, 2005.

Baur, Dorothea. *NGOs as Legitimate Partners of Corporations: A Political Conceptualization*. New York: Springer, 2011.

Becchetti, Leonardo, and Carlo Borzaga, eds. *The Economics of Social Responsibility: The World of Social Enterprises*. London: Routledge, 2010.

Beecroft, Adrian. *Report on Employment Law*. United Kingdom Department for Innovation and Skills, October 24, 2011.

Berkowitz, Jeremy, and Michelle J. White. "Bankruptcy and Small Firms' Access to Credit." *RAND Journal of Economics* (Spring 2004): 69–84.

Bhattacharya, C.B., Sankar Sen, and Daniel Korschun. *Leveraging Corporate Responsibility: The Stakeholder Route to Maximizing Business and Social Value*. Cambridge: Cambridge University Press, 2011.

"BPIFrance." *BPIFrance*. www.oseo.fr

Brockner, Joel. "Why It's So Hard to be Fair." *Harvard Business Review* 84, no. 3 (2006): 122–29.

Collinson, David, Keith Grint, and Brad Jackson, eds. *Leadership*. Thousand Oaks, CA: Sage Publishing, 2011.

"Companies and Productivity: Small Is Not Beautiful." *The Economist*, March 3, 2012.

DeMartino, George F. *The Economist's Oath: On the Need for and Content of Professional Economic Ethics*. Oxford: Oxford University Press, 2010.

Dutta, Soumitra, ed. *The Global Innovation Index 2012: Stronger Innovation Linkages for Global Growth*. INSEAD and World Intellectual Property Organization, 2012.

European Commission. *Action Plan: The European Agenda for Entrepreneurship*. February 11, 2004. Brussels, European Commission.————. *Minimizing Regulatory Burden for SMEs*. November 23, 2011. Brussels, European Commission.

————. *A Second Chance for Entrepreneurs: Prevention of Bankruptcy, Simplification of Bankruptcy Procedures and Support for a Fresh Start*. Final Report of the Expert Group, January 2011. Brussels, European Commission.

"Explaining Economic Weakness: The IMF v Beecroft." *The Economist*, May 26, 2012.

Fan, Wei, and Michelle J. White. "Personal Bankruptcy and the Level of Entrepreneurial Activity." *Journal of Law and Economics* (October 2003): 543–67.

Freeman, R. Edward. *Strategic Management: A Stakeholder Approach*. Boston: Pitman Publishing, 1984.

Freeman, R. Edward, and John McVea. *A Stakeholder Approach to Strategic Management*. Darden Business School Working Paper No. 01–02, 2001.

Hamel, Gary. "First, Let's Fire All the Managers." Harvard Business Review (December issue 2011): 49–60. https://hbr.org/2011/12/first-lets-fire-all-the-managers

Hamschmidt, Jost, and Michael Pirson, eds. *Case Studies in Social Entrepreneurship and Sustainability*. Sheffield, UK: Greenleaf Publishing, 2011.

Hitt, Michael A., R. Edward Freeman, and Jeffrey S. Harrison, eds. *The Blackwell Handbook of Strategic Management*. Blackwell Publishing, 2006.

IBM. *Global Innovation Outlook 2.0*. March 2006.

International Finance Corporation. "IFC SME Ventures Fund." www.ifc.org

International Monetary Fund. *Arab Countries in Transition: An Update on Economic Outlook and Key Challenges*. Washington, DC: International Monetary Fund, 2014. http://www.imf.org/external/np/pp/eng/2014/040914.pdf

Judah, Tim. "Senegal's Mourides: Islam's Mystical Entrepreneurs." *BBC News*, August 4, 2011.

Kane, Tim. *The Importance of Startups in Job Creation and Job Destruction*. Kauffman Foundation of Entrepreneurship, July 2010. www.kauffman.org

Kelley, Donna J., Slavica Singer, and Mike Herrington. *Global Entrepreneurship Monitor: 2011 Global Report*. 2011. www.gemconsortium.org

Landes, David, Joel Mokyr, and William Baumol. *The Invention of Enterprise: Entrepreneurship from Ancient Mesopotamia to Modern Times*. Princeton: Princeton University Press, 2010.

Liechti, Diego Dimitri. *Determinants of Entrepreneurial Activity and Success*. Geneva: Swiss Private Equity & Corporate Finance Association, March 2011.

"Life after Debt." *The Economist*. April 16, 2005.

Mikdashi, Zuhayr. *Progress-driven Entrepreneurs, Private Equity Finance and Regulatory Issues*. Basingstoke, UK: Palgrave Macmillan, 2010.

Molian, David. "Entrepreneurial Value Creation: Are European Business Schools Playing Their Full Part?" Paper presented at the European Foundation for Management Development Conference, Maastricht, March 5–6, 2012.

"The Morning Star Company." *Morning Star*. http://www.morningstarco.com.

OECD. *SMEs, Entrepreneurship and Innovation*. June 2010.

Oxford Centre for Entrepreneurship and Innovation's Venture Club, Saïd Business School. www.sbs.ox.ac.uk

Post, James E., Lee E. Preston, and Sybille Sachs. *Redefining the Corporation: Stakeholder Management and Organizational Wealth*. Stanford: Stanford University Press, 2002.

République Française—Direction du commerce, de l'artisanat, des services et des professions liberals. *Statut de la PME de Croissance—guide pratique*. Paris, 2007. http://www.bnf.fr/documents/biblio_tpe.pdf

Schumpeter, Joseph. *Capitalism, Socialism, and Democracy*. New York: Harper & Row, 1950.

Skelton, Rose. "Senegal Clothes Maker Gives Faithful Jobs and Hope." *BBC News*, June 20, 2006. http://news.bbc.co.uk/2/hi/africa/5075534.stm

"Startup America Partnership." Startup America Partnership, LLC. http://www.s.co.

"The Trouble with Small Firms." *The Economist*. March 3, 2012.

UN High Commission for Refugees. "Mediterranean Takes Record as Most Deadly Stretch of Water for Refugees and Migrants in 2011." *Brief Notes*, January 31, 2012.

UNU-IHDP and UNEP. *Inclusive Wealth Report 2012: Measuring Progress toward Sustainability*. Cambridge: Cambridge University Press, 2012.

Vandenberg, Paul. "Poverty Reduction through Small Enterprises." Working Paper. Geneva: International Labour Office, 2006.

"Visas for Entrepreneurs: Where Creators Are Welcome." *The Economist*, June 9, 2012.

World Bank. *Doing Business 2012*. www.doingbusiness.org

World Bank. *MENA Quarterly Economic Brief*. Washington, DC: World Bank, 2014. http://www.worldbank.org/content/dam/Worldbank/document/MNA/QEBissue2January2014FINAL.pdf

Zeitoun, Hossam. "Corporate Governance Modes, Stakeholder Relations, and Organizational Value Creation." PhD diss., University of Zurich, 2011.

2. Facilitating Entrepreneurship as a Catalyst for Change

Shailendra Vyakarnam and Shima Barakat

Entrepreneurship is becoming a social movement, and is also seen as a catalyst for change across much of the Middle East. This chapter focuses on lessons that have been learned in the process of shaping graduate-level entrepreneurship education. Such lessons inform institutions of changes needed at macro, meso, and micro levels. Many programs and projects in entrepreneurship development are already underway, and in such cases the focus will need to be on scaling with some urgency. For those who are at the early stages of entrepreneurship development, it is hoped this chapter provides a framework for future plans.

The Context of the Challenge

Since the beginning of 2011, dramatic changes have taken place in much of the Middle East and North Africa. To understand the impacts of such societal transformations on people's livelihoods, we need to look back to the fall of the Iron Curtain in 1989.[1] Prior to 1989, there were broadly three forms of labor economy: one that operated under socialism with planned production and strictly allocated resources; one that operated under free market conditions; and one where markets were moderated by a welfare state, often combined with nationalist industrial policies.

Broadly, western free-market capitalism and the socialist planned economic system each had workforces approximating 1.5 billion prior to 1989, but the strict political, societal, and economic separation of the two systems meant that these workforces did not compete with each other. After 1989, socialist and welfare-based economies transitioned into an ever more deregulated 'free market' capitalist system in which over three billion workers now compete.[2] The impact of this seismic shift on resource allocation, employment opportunities, migration patterns, human and intellectual capital competing in one marketplace, the rate of globalization, and so forth cannot be underestimated. It is in this wider socioeconomic context that we need to understand the relevance of entrepreneurship in the Middle East and, more specifically, Egypt today.

Why Is This Important?

With more than 50 percent of the population in the Arab world under the age of 25,[3] combined with the societal forces of the "Arab Spring," demographics in the region have created a potential workforce that will be pushed into an already oversupplied global labor market. This force is late in joining and, due to its recent political history, may not be suitably equipped to compete with peers from countries that have had twenty years to adjust to global economic changes. Further competitive pressure for the estimated 85 million jobs that are needed in the Middle East and North Africa over the next ten years comes from the growing democratization of its African neighbors, stagnant European economies, the demographic (youth) dividend of India, and the ticking time bomb of China's "one child per family" policy. This means there is urgent need to find ways to turn large pools of unengaged labor into both producers and consumers of new products and services. Ignoring this urgency and the sheer numbers could be disastrous.

The notion of self-employment and of job creators rather than job seekers that has been raised in policy circles in reaction to this context is not new. In fact, the very early definition of 'entrepreneur' as one who takes risks because of the uncertainties associated with self-employment (rather than remaining a non-risk-taking job seeker) was framed as far back as the nineteenth century. Later, M.K. Gandhi, caricatured as a man in a loincloth weaving his own fabric in the name of self-sufficiency, also pursued the idea of micro-enterprise and self-sustainability to rid India of its dependency on Britain. In the 1970s, the economist E.F. Schumacher coined a famous phrase, "small is beautiful."[4] His book argued that labor dignity and personal freedom come from working for oneself rather than being dependent on large corporations for income. Schumacher was flying in the face of J.K. Galbraith's now infamous assertion that the era of small businesses was over and that all future business would be conducted by large corporations with the capacity to scale. For most of the twentieth century, and particularly since the early 1980s, the image of the entrepreneur that has dominated the public sphere is one of individual heroism: a dominant, masculine, and western character who pursues his own vision relentlessly, bringing about commercial renewal and accumulating great personal wealth in the process.[5]

Over the past fifteen years, the dominant narrative on entrepreneurship has come to explicitly include the notion that entrepreneurship is a catalyst for change, specifically in scientific enterprise and social enterprise.[6] Entrepreneurs are said to be particularly well-placed as potential change agents. Business and enterprise also have an increasingly prominent role in taking responsibility for managing social and environmental externalities, and even to "create shared value."[7] In this context, we now have an increasingly improved understanding of the potential of entrepreneurship, even if it carries different names, such as 'inclusive growth,' 'microenterprise,' and 'markets at the bottom of the pyramid.'

Our starting point is to accept the notion of entrepreneurship as a catalyst for change, and to recognize its transformational potential in the social, economic, political, and environmental realities of the Arab world of today. To capitalize on this potential, it is essential to consider the appropriate competencies, frameworks, and policy environments that are required to make a genuine difference to millions of young, aspiring, creative, and hopeful people. More specifically, how and what can universities and government institutions do to foster entrepreneurship? This chapter attempts to summarize these issues in order to assist with policy making, culture change, and generating short-term impact on young graduates, chosen here to act as role models for society in the unleashing of their educated talent to solve today's problems.

For entrepreneurship to facilitate change, numerous issues need to be resolved. For us, these begin with providing practitioner-led entrepreneurship education. This is a distinct approach when contrasted with much of education in universities that is "about" entrepreneurship. To bring about change through this practice-based model, there are also some preconditions in policy issues that need to be addressed. In what follows, we draw on lessons learned at the University of Cambridge when this very traditional university was tasked with shifting its culture toward a more positive, enterprise-friendly environment in order to bring about change. We conclude that focusing on entrepreneurial learning is essential in equipping individual entrepreneurs with the skills to succeed in creating economic, social, and environmental value and to innovate and engineer genuinely transformative solutions.

Entrepreneurship Development

Analytically, it is useful to separate entrepreneurship development into three layers. First is the macro level, which addresses overall policies that create a conducive atmosphere for entrepreneurship. Second is the meso layer, which includes programs that are thematic and provide funding and levers for making culture change a reality. Third is the micro level, which is for projects, courses, incubation services, training, and education to provide the skills to deliver the meso and macro changes needed.

Macro-level policies

It is beyond the scope of this chapter to examine the broad policy landscape that influences a supportive climate for entrepreneurship; the Arab/Middle Eastern context is particular and rapidly changing, which warrants a separate study. However, some of the policy issues that need to be addressed by governments wishing to promote enterprise are illustrated in table 2.1.

Table 2.1 Frequently-cited policy changes that enable entrepreneurship development

Reduce the amount of red tape needed to start businesses. Some highly streamlined countries enable new ventures to register and get underway in a matter of a few days, including all the formal registrations, setting up of bank accounts, and flexible terms for renting space.
From a government perspective, streamlining a registration process typically means 'single window' approaches where startups can be accelerated through the various procedures.
Ease tax burdens and fiscal regulations. It is important to simplify taxation in order to obtain better compliance. A burdensome system means that entrepreneurs who are already struggling will neglect this important element of their duties to society.
Implement intellectual property rules that are simple but effective. There is often a misunderstanding about intellectual property, whether a specific patent or, more generally, copyrights, brands, and the ability to secure a unique position for one's products and services to compete effectively in the market. A transparent legal system based on awareness and education is key to this aspect of entrepreneurship development.
Encourage banks to lend to the SME sector by helping mitigate banks' risks in this sector. Government intervention is essential for this to be a successful policy measure.

Facilitate easy employment rules for owners of small businesses. Employment creation is one of the positive outcomes of a successful enterprise culture. But to get momentum, employers must have sufficient flexibility to make commercial decisions as well as provide security with good employment terms. A few simple principles should guide fair terms and conditions.

Simplify business closure—make it acceptable for businesses to fail.[1]

Such measures "de-criminalize" entrepreneurship or, put in a more positive way, they are needed to liberalize the environment so that people feel free to start companies.[1]

Meso-level activities

The primary function of activities at the meso level is to effect culture change. There is talk about "entrepreneurial universities,"[2] but that may be going one step too far. It could be sufficient that universities become friendly toward enterprise/entrepreneurship and take a more lenient view of the efforts of faculty and students. This could be achieved by being involved in and hosting open lecture series, competitions, and campaign celebrations, all of which combine to create a culture that is positive toward entrepreneurship.[3] It is at this level that medium-term funding can be applied in order to allow deep-rooted programs to have impact.[4]

Most important among the activities to be developed at the meso level is for the heads of institutions to signal the importance of entrepreneurship, not only through strategy statements but also by backing the agenda with a tangible provision of resources. They need to signal this importance to faculty members, establishing career rewards for teaching entrepreneurship development. Universities may find this hard to implement and develop, as many are burdened by a heavy legacy of past practices and with a culture that emphasizes governance rather than outcomes. They need to consider and decide on the scope of entrepreneurship education, the policies that support entrepreneurship, the resources they are willing and able to allocate, the scale of engagement and impact, curriculum development, and finally the capacity to deliver.

Scope

Universities need to decide on the nature and scope of entrepreneurship education. Issues to consider include:

- Is it for employability or self-employment?
- Is it to be mandatory for all graduates, or an option?
- Will it be available across all disciplines?
- Is it seen as the domain of business schools or as a university-wide activity?

These questions and discussions lead to clarity on organizational arrangements for the delivery of entrepreneurship development at a university. They are political in nature, and can be driven by existing structures and forms, but it is discussion around scope that will help bring about shared values and vision for delivering entrepreneurship. In their programs' values, institutions need to find ways to encourage diversity, multi-disciplinary interactions, and gender sensitivity to best harness talent in open conversations that create opportunities for entrepreneurship.

Policies

A university's policies will need to be reviewed and often reorganized to encourage entrepreneurship. These include employment terms and conditions for faculty that provide freedom to operate in the event that they wish to engage in new venture creation. In addition to reviewing employment terms, universities need to initiate clear intellectual property (IP) rules that are fair and in favor of commercialization. Talented and empowered staff members are also needed at technology-transfer offices dealing with IP if the policies are to run smoothly.

In order to encourage student enterprise, faculty must find new ways to provide assessment, develop curricula, and be free to engage in pedagogical innovation. The use of local entrepreneurs and alumni as external lecturers can provide inspiration, talks that affect mindset, and highly practical advice. Such powerful inputs can achieve culture change. To trigger this inclination in practitioners, it is necessary to make available physical infrastructure for talks, events, and networking sessions. These should be accompanied by hospitality. But underpinning all this activity, faculty must be committed and understand content and process.

Resources

The changes needed at the meso level require resources, ideally with core funding that is stable over time so that programs of activity have time to take root. Funding should develop and deliver new courses, as well as cover expenses for entrepreneurs and other professionals involved in them. Funding must enable a reaching out to 15 to 20 percent of students for the "cultural needle" to shift in favor of entrepreneurship as a career option.

In the longer term it may be necessary to provide physical resources for entrepreneurship development, such as lecture theaters and open spaces for networking and hospitality. They may also include incubation spaces, shared facilities for engineering prototypes or wet labs for biotechnology, media labs for creative industries, and design schools for product and service design. These are somewhat dependent on the nature of each university and the core competences from which it will develop its entrepreneurship programs.

Scale

Instituting culture change in favor of entrepreneurship at universities requires scalability. When new courses are launched or developed, due to resource constraints they are often run for relatively small numbers of students. This small-scale approach will not alter the mindsets of young graduates, as students who participate in courses with only a few participants will feel like outliers. To make entrepreneurship a more acceptable future for young people, the activities instituted by universities must be large in ambition and delivery.

Curriculum development

Universities need to incorporate entrepreneurship into their curriculums. We provide an example of developing a portfolio of courses relevant to audiences that are attempting to answer basic questions related to entrepreneurship. The curriculum has been formed around the responses to these questions, which are:

- What is entrepreneurship?
- Is it for me?
- When is an idea a business opportunity?
- I have an idea—what should I do next?

Our experience with curriculum development around these questions has been positive and allows us to recommend that institutions seeking to initiate entrepreneurship programs determine the questions that are most relevant to their audiences, and help answer them.

Capacity for delivery

This may be one of the most difficult challenges. The direction of capacity development should be based on an institution's scale, scope, and curriculum direction. In our interactions with several institutions, we have seen a number of strategies that we believe are mistaken if the purpose is to effect culture change. These errors include:

- Double appointments of senior faculty members in entrepreneurship and another unrelated discipline. This results in faculty members being torn between their core discipline and entrepreneurship, which they feel obliged to take on as a burden and without fully understanding what is required of them. Often faculty are denied resources to properly update themselves through research or conference attendance. Because they are not embedded in the topic, they may well lack rigor and may not have access to relevant case studies. This has a cumulative effect in creating a distant faculty who do not carry the students with them on the topic.
- Giving entrepreneurship high visibility to attract student numbers but failing to follow through at the micro level, with budgets, resources, and faculty that are fully engaged, because of weak policies and strategies.
- Young faculty appointed in entrepreneurship. This can be intimidating because while their careers are dependent on publications, heavy teaching loads in entrepreneurship can sometimes deny them the time and resources needed to publish. A separate but related problem is that entrepreneurship journals have yet to make it into the big league, so talented young faculty drift away from the field to secure their careers by publishing in mainstream management journals.

Senior university members need to review their position on these and other challenges to capacity for delivery and help to overcome them if there is to be real progress at the meso level.

Micro-level activities

Micro-level delivery of initiatives is essential to achieve the objective of meso-level initiation of culture change. There are a number of philosophical debates that institutions first need to engage in, not least to develop definitions of entrepreneurship that are relevant to their constituent groups. For example, what enterprise size is considered entrepreneurship? In Silicon Valley, entrepreneurship means starting companies like Google. In Europe, it may mean starting firms that can grow to around 250 employees (so-called SMEs). Elsewhere, it can mean self-employment or even microenterprise creation using self-help groups and microfinance mechanisms.

It is essential therefore to be clear about target groups and desired outcomes. Beyond contextual considerations, we also need to account for age groups, socioeconomic groups, levels of education, and types of capability that already exist in sectors. Starting a tourism venture when there is nothing to see or aspire to, for example, or an Internet company when there is little or nothing in the realm of computer studies available, demands a certain level of strategic analysis.

If we continue with our assumption that long-term cultural change will come about through the education of university graduates who not only start ventures but also begin to occupy jobs in policy making and managerial roles in companies and even transfer their know-how to family and friends, we can take a deeper look at what kinds of curricula would be most appropriate.

Education for Entrepreneurship

Translating meso-level considerations to micro-level delivery of programs requires a clear pedagogical framework, as we learned in our experience at Cambridge, where we chose to address this as *practitioner-led education for entrepreneurship*. The primary reason for this view is that entrepreneurship can be considered a profession like medicine, law, or dentistry. Therefore, students need to appreciate theory but also have deep confidence and capability in practice. In our view, the best people to deliver an education that will enable students to become entrepreneurs are practitioners who have expertise and experience in entrepreneurship. The faculty's clear-cut role is to develop appropriate curricula and create a learning atmosphere, but without the active engagement of practitioners such as entrepreneurs, who are invited and carefully briefed, it is very difficult for students to feel they are gaining real, practical knowledge that will enable them to actually go out and create new ventures or become more employable. The core differences between curriculums designed to educate students *about* and *for* entrepreneurship are listed in table 2.2 below.

Table 2.2. Key differences between "about" and "for" education for entrepreneurship

About	For
Case studies	Role models in class
Lectures about theory and SME discussions	Effectuation
Simulation	Learning by doing
Videos	Venture creation
Essays	Internships
Exams	Project assessment blending theory with practice
Teaching by academics without practice	Teaching by practitioners—with curriculum

Defining Entrepreneurial Learning

In order to begin to develop a suite of programs, activities, and events, we need to better understand what is meant by entrepreneurial learning in support of culture change. This can be defined as the need to integrate three components of learning (see fig. 2.1):

- Personal awareness: Do individuals understand what it means to be an entrepreneur, and do they have the desire and ability to pursue such a career? This entails understanding their ethical position, values, and motivations.
- Business know-how: Those who wish to enter entrepreneurial careers in business actually do need to know the basics of business. They need to understand market needs, pricing, marketing, commercial terms, finance and operational issues, and so on.
- Social skills: This includes team building, presenting, selling, negotiating, being presentable, body language, closing the sale, and ensuring delivery of promises.

Often these elements of entrepreneurial learning are separated, or only a small part of the learning is provided, such as business know-how or bookkeeping and marketing/selling. Educators need to frame programs or activities in more holistic ways.

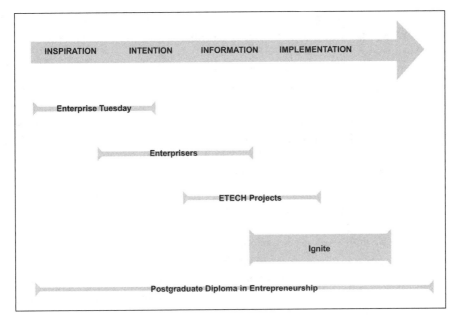

INSPIRATION   INTENTION   INFORMATION   IMPLEMENTATION

Enterprise Tuesday

Enterprisers

ETECH Projects

Ignite

Postgraduate Diploma in Entrepreneurship

Figure 2.1 Entrepreneurial learning model

We see the learning process as a journey, from wanting to raise one's awareness to seeking help and support in creating ventures. One way of looking at this journey is to conceptualize it around the value chain.[5] The journey begins with inspiration, continues with raising self-confidence and intention to form ventures (personal awareness), progresses toward securing "how-to" information (business know-how), and is completed by assisting with implementation (practice with social skills).[6]

Allied to these learning objectives and the journey-taking approach, micro-level programs also need to be embedded in the local ecosystem of support. The primary reason for this is that the teaching and venture-creation parts may need to be separated in order to avoid issues of conflicts of interest, resource deployment, and continuing ability to scale.

For a vibrant community to develop based on entrepreneurship as a catalyst for change, universities need to have micro-level initiatives that bring together the wider business community in informal ways to provide a full set of interactions between industrialists, entrepreneurs, academics, students, investors, and university departments. It is through these multidisciplinary activities and conversations that opportunities open up. Additionally, confidence is gained in such environments for new venture creation by young graduates who feel there is a safety net in the very early stages of their experiments with entrepreneurship as a future career direction.

For example, as part of the efforts of the Centre for Entrepreneurial Learning at Cambridge to deliver a practitioner-led curriculum, the authors have created an active set of linkages to foster an entrepreneurial ecosystem. This ecosystem, through the linkages established, allows for the delivery of optional courses for students. Links are made with student societies, the local business community, technology-transfer offices, and numerous departments within the university. Accordingly, students who have a predetermined idea of what to study, whether it is engineering or sciences or arts and humanities, are exposed to entrepreneurship education.

Through working in collaboration with the various members of the ecosystem, students see the world in more multidisciplinary ways. They engage more swiftly with people from other departments and the business community, and connect with people and institutions that can help them once they are ready to go into the wider world.

Measuring Outputs and Outcomes

Measurement of effort is crucial in the area of entrepreneurship because expected outputs include culture change, real ventures, and the general improvement of social well-being. For some reason, the same demands are not made in other disciplines in academia, such as economics or business studies, which are seen as general studies that raise capability levels. Entrepreneurship education is seen as a profession that is expected to deliver results.

What metrics does one need to take into account? One line of argument would suggest measuring the outcomes of the learning objectives of a given course. But it becomes harder to set out fixed measures when considering the development of ecosystems, the assistance of alumni in teaching, and the creation of events and activities that stimulate enterprise creation.

Three overall forms of measurement have been created. These measure the outcomes associated with specific courses, contributions to venture creation more generally, and, more recently, changes in mindset in terms of self-efficacy,[7] using instruments that are based on a before-and-after approach to data gathering.

Self-efficacy is defined as a person's confidence in their own ability to successfully carry out a task or activity. Entrepreneurial self-efficacy relates to an individual's confidence in being able to successfully perform the tasks and activities associated specifically with being an entrepreneur. It is identified as an important variable in determining not only the strength of entrepreneurial intentions but the probability that such intentions will lead to actions.[8]

It can therefore be argued that entrepreneurial self-efficacy can and should be recognized as a valuable program objective and, similarly, can be used as an efficient and effective measure of program/course outcome and impact. It is particularly valuable in that its predictive power means that outcomes can be measured years before activities like venture creation can be recorded.

Discussion

As a potential catalyst for technological and social change, entrepreneurship has emerged center-stage in policy and academia in various political, economic, and social contexts around the globe over the past fifteen years. While the cultural connotations of enterprise, risk, and entrepreneurial selfhood vary greatly from region to region, many societies seem at the very least to be gradually shifting from a mindset of suspicion to one of curiosity about entrepreneurship. As a result of this increased level of activity and interest, many institutions around the world have started to engage seriously with entrepreneurship, as acknowledged in the World Economic Forum publication[9] and the recent book on Unlocking the Enterpriser Inside.[10] A growing body of evidence (of SME policy, educational interventions, and investments) suggests that improvements in the level of education for entrepreneurship, provision of infrastructure, incentives, and incubation services help societies recover from financial and political crises.

In this context, we argue that governments and universities in the Arab world need to take stock of their macro, meso, and micro policies, programs, and actions. For a country such as Egypt, with its unique mix of nationalist industrial policies, elaborate bureaucratic systems and regulatory regimes, and a rapidly fluctuating economic climate, it is clear that creating an attractive environment for entrepreneurship presents many challenges. Yet rapid urbanization, a young demographic, and the fragmented provision of essential services also creates tremendous opportunities for innovative solutions.

The differing timescales of long-term policy change and immediate societal challenges create major tensions around the question of how to equip aspiring entrepreneurs with the skills to innovate.

Providing educational interventions for entrepreneurship will thus, at least in the short-term, need to focus on the development of an entrepreneurial mindset to create the conditions for graduates to develop greater awareness of opportunities. These interventions could be provided in parallel with tailored programs and incubator spaces that recognize the specific institutional context and the societal challenges so as to enable the growth in numbers of entrepreneurs. There is a challenge at this point for the educators, who may themselves need to recognize their own mindsets and skills in being able to impart the inspiration and the knowledge needed by their graduates in fostering entrepreneurship.

Industry is a third party to the alliance between governments and universities. University alumni, business managers, industrialists, and others are contributing in positive ways by bringing in experience and context, which strengthens policies, education, and the realization of opportunities. Certain clusters, such as those around MIT, Stanford, Cambridge, Oxford, Imperial College London, Indian Institutes of Technology, and the American University in Cairo, have been most effective in leveraging their entrepreneurial alumni. This chapter has not dwelled on interactions with industry, but this is absolutely essential if progress is to be made. The world of practice is so far removed from the world of study that the sooner connections are made the better, especially if we are expecting students to go into society to make a difference through entrepreneurship.

Universities have mechanisms of governance and academic requirements that often prevent this kind of interaction. These are becoming outdated modes of academic administration and need to be reviewed. We have found that entrepreneurs value their interaction with universities and both the formal and informal recognition they receive from it, to the extent that they will often not charge for teaching or mentoring sessions. Much work has to be done on the operational details of such arrangements, but if we agree that entrepreneurship can be a catalyst for change, we have to find solutions to these challenges. The more we enable and empower such change, the more we can rely on it to impact society.

In conclusion, separating education *about* entrepreneurship from education *for* entrepreneurship can be a basis for macro policies, meso-level culture change, and micro-level actions that provide immediate results for students and society. This, we argue, is key to making a genuine impact on the livelihoods of millions of young individuals in a rapidly transforming Middle East.

Notes

1 Note the impact that freedom to act has had on India and China—which were both previously highly regulated environments.

2 See recent writings by Professor Alan Gibb, such as Paul Coyle, Alan Gibb, and Gay Haskins, *Creating the Entrepreneurial University: From Concept into Practice and Impact* (Coventry: Entrepreneurship in Education, 2013); Alan Gibb and Gay Haskins, *University of the Future: An Entrepreneurial Stakeholder Learning Organisation?* (Coventry: Entrepreneurship in Education, 2013); Alan Gibb, *The University Entrepreneurial Scorecard: Reviewing the Entrepreneurial Potential of a University* (Coventry: Entrepreneurship in Education, 2013).

3 Examples include: Students in Free Enterprise; Global Entrepreneurship Week; business plan competitions; and various projects run by multinationals such as Microsoft, Intel, Cisco, Santander Bank, and so on.

4 We have seen medium-term funding in the United Kingdom with the Higher Education Innovation Fund (HEIF). There have now been five rounds of HEIF, each of two years' duration and each attracting GBP200 million, distributed across most of the 100 plus universities. HEIF funding has enabled numerous activities, events, education programs, incubation, technology transfer offices, competitions, networking activities, and conferences to emerge as part of a growing landscape of activity. These have become university-wide initiatives. There is an increasing number of professorships, conferences, journal articles, and case studies. All these efforts are exemplary in terms of building a critical mass of lecturers and facilitators. One outcome of HEIF funding is the formation of a member-led organization (Enterprise Educators UK) that has a strong ethos of sharing materials, methods, and tacit knowledge in entrepreneurship education.

5 Based on Michael E. Porter, *Competitive Advantage of Nations* (New York: The Free Press, 1990).

6 "Centre for Entrepreneurial Learning," *University of Cambridge Judge Business School,* http://www.cfel.jbs.cam.ac.uk/.

7 For a deeper discussion on self-efficacy as a measure of entrepreneurial intent, see research papers by Dr. Shima Barakat, such as: Monique Boddington and Shima Barakat, "Measuring Creative Learning Activities: A Methodological Guide to the Many Pitfalls," *Interdisciplinary Studies Journal* 2 (2013): 195–206; Becky Schutt et al., "A Smooth Sea Never Made a Skilled Sailor: DERO Project: Research Findings and Insights" (Report to NESTA and AHRC, 2013).

8 See for example: Chao C. Chen, P.G. Greene, and A. Crick, "Does Entrepreneurial Self-efficacy Distinguish Entrepreneurs from Managers?" *Journal of Business Venturing* 13 (1998): 295–316; Alexandra L. Anna et al., "Women Business Owners in Traditional and Non-traditional Industries," *Journal of Business Venturing* 15 (2000): 279–303; Robert J. Baum and E.A. Locke, "The Relationship of Entrepreneurial Traits, Skill and Motivation to Subsequent Venture Growth," *Journal of Applied Psychology* 19 (2004): 587–98.

9 Christine Volkmann et al., "Educating the Next Wave of Entrepreneurs: Unlocking Entrepreneurial Capabilities to Meet the Global Challenges of the 21st Century," in *World Economic Forum: A Report of the Global Education Initiative*, Dianna Rienstra and Nancy Tranchet, eds. (Geneva: World Economic Forum, 2009).

10 Neal Hartman and Shailendra Vyakarnam, *Unlocking the Enterpriser Inside! A Book of What, Where and How!* (London: World Scientific, 2011).

Bibliography

Anna, Alexandra L., G.N. Chandler, E. Jansen, and N.P. Mero. "Women Business Owners in Traditional and Non-traditional Industries." *Journal of Business Venturing* 15 (2000): 279–303.

Baum, J. Robert, and E.A. Locke. "The Relationship of Entrepreneurial Traits, Skill and Motivation to Subsequent Venture Growth." *Journal of Applied Psychology* 19 (2004): 587–98.

Boddington, Monique, and Shima Barakat. "Measuring Creative Learning Activities: A Methodological Guide to the Many Pitfalls." *Interdisciplinary Studies Journal* 2 (2013): 195–206.

Chell, Elisabeth. "Social Enterprise and Entrepreneurship: Towards a Convergent Theory of the Entrepreneurial Process." *International Small Business Journal* 25 (2007): 5–26.

Chen, Chao C., P.G. Greene, and A. Crick. "Does Entrepreneurial Self-efficacy Distinguish Entrepreneurs from Managers?" *Journal of Business Venturing* 13 (1998): 295–316.

Coyle, Paul, Alan Gibb, and Gay Haskins. *Creating the Entrepreneurial University: From Concept into Practice and Impact.* Coventry: Entrepreneurship in Education, 2013.

Gibb, Alan. *The University Entrepreneurial Scorecard: Reviewing the Entrepreneurial Potential of a University.* Coventry: Entrepreneurship in Education, 2013.

Gibb, Alan, and Gay Haskins. *University of the Future: An Entrepreneurial Stakeholder Learning Organisation?* Coventry: Entrepreneurship in Education, 2013.

Hartman, Neal, and Shailendra Vyakarnam. *Unlocking the Enterpriser Inside! A Book of What, Where and How!* London: World Scientific, 2011.

Kramer, Mark, and Michael E. Porter. "Creating Shared Value." *Harvard Business Review* 89, no. 1/2 (2011): 62–77.

Mirkin, Barry. *Population Levels, Trends and Policies in the Arab Region: Challenges and Opportunities.* United Nations Development Program. Regional Bureau for Arab States: Arab Human Development Report, 2010.

Nicholls, Alex, ed. *Social Entrepreneurship: New Models of Sustainable Social Change.* Oxford: Oxford University Press, 2006.

Ogbor, John O. "Mythicizing and Reification in Entrepreneurial Discourse: Ideology-Critique of Entrepreneurial Studies." *Journal of Management Studies* 37 (2000): 605–35.

Porter, Michael E. *Competitive Advantage of Nations.* New York: The Free Press, 1990.

Schumacher, Ernst Friedrich. *Small Is Beautiful: A Study of Economics As If People Mattered.* London: Blond and Briggs, 1973.

Schutt, Becky, Matthew Petrie, Allègre L. Hadida, Shima Barakat, and Adrian B. Cruz. "A Smooth Sea Never Made a Skilled Sailor: DERO Project: Research Findings and Insights." Report to NESTA and AHRC, 2013.

University of Cambridge Judge Business School. "Centre for Entrepreneurial Learning." http://www.cfel.jbs.cam.ac.uk/

3. Putting the Horse before the Cart: Understanding Creativity and Enterprising Behaviors

Andy Penaluna and Kathryn Penaluna

Introduction

Historically, there has been great emphasis on the development of business plans in educational programs relating to enterprise and entrepreneurship, and most assessment strategies aiming to evaluate student performance have focused on them.[1] In most business school environments, the assumption is that the learner always brings along an idea to develop, and that this idea will become the focal point for a group or individual project. In these scenarios, the development of a good idea is often tacitly presupposed or receives little attention.

In contradistinction to a business plan developed in an educational context, "bootstrapping" is based on the concept that a young business will start with nothing of great value and is reliant on the skillset and innovative capacity of the entrepreneur.[2] As a strategy, it often conflicts with what venture capitalists require, because financial support tends to come from friends and family, as opposed to more traditional sources that anticipate a business plan. It also deviates from what venture capitalists require because the strategy tends to empower the idea generator to take things further in a series of incremental steps and to make strategic short-term plans that can easily shift and change, as opposed to locking in the proposal in a fixed way. Another and not too dissimilar alternative perspective is Eric Ries's "lean startup,"[3] where the manufacturing methodology of being able to shift and change, usually according to customers' needs, drives the process and continuously redefines the plan.

Developing the initial business idea remains a key factor, as without it no further action is possible. Moreover, if lean or bootstrap strategies are employed, it is the entrepreneur or entrepreneurial team's ability to continuously spot opportunities and to adapt the shifting circumstances that becomes the paramount skill requirement. These are the domains of creative and innovative thought, and we shall argue that these skills should be better considered and more appropriately situated within learning.

In an educational setting, it is normal for the idea to be rigorously examined, tested, and evaluated against the predictions of a fixed plan, not only to see that it has the potential to be the seed of a new business venture but also to ensure that the assessment of the learning can be facilitated within fixed timescales such as terms, semesters, or even years. Fiscal evaluations are often at the core of this type of student assessment, and help the educator to determine the potential success of the proposed venture. If the educator sets long-term goals that require rigid outcomes, however, this fixed approach ignores aspects such as the rapidly changing marketplace and the determination, innovative ability, and resilience of the entrepreneur or entrepreneurial team. Where, we ask, can the student demonstrate the ability to respond to constant change?

Moreover, in our educational settings, adaptions and changes are often suggested in response to the student business plan, but the initial generation of varied ideas and multiple alternatives, which in turn facilitate flexibility and change, is rarely considered beyond short brainstorming or problem-solving exercises.

It follows that the experience of having to shift positions rapidly and creatively in response to changing environments and situations, and to be assessed on the associated abilities to adapt, adjust, and reformulate, require learning environments that are continuously customer feedback-oriented, flexible, adaptable, and context-driven.

Creativity specialists know that short-term approaches offer limited results, as creativity is reliant on behaviors such as challenging norms and seeing alternative perspectives. As behavior takes time to change and develop, short-term creativity events will most likely offer few long-term results. In order to behave differently, for example to be comfortable in situations of ambiguity and risk, longer-term learning strategies that encompass this understanding are needed. We will therefore argue that an understanding of behavioral change is especially relevant to entrepreneurial education, yet it is largely ignored in the context of a business plan assessment that only focuses on fixed long-term predictions.

Background: The Development of Teacher Training for Entrepreneurship

In 2010, the Welsh government funded a year-long feasibility study on the potential for a formally approved teacher-training qualification in enterprise and entrepreneurship. It became apparent during the study that the UK provision for teaching educators was only being offered as continuous professional development, which is to say that it was unstructured and mainly ad hoc. No formally recognized university-level module or unit was available in UK higher education. This was acknowledged to be an important omission, because robust university course validation procedures ensure quality and standing and are subject to considerable scrutiny prior to approval, without which standards of achievement were suspect at best.

As a more sustainable approach to embedding enterprise and entrepreneurship in all schools and colleges was a primary government goal, a totally new course of teacher study had to be conceptualized and developed. The UK team tasked with this role was aware that formal approval within university validation procedures would be a significant hurdle, not least because with no existing benchmarks to follow, new arguments would have to be made and new research undertaken. Therefore a robust research exercise was initiated to investigate what would be needed to assure the quality and standing of a new and innovative entrepreneurial educator course.[4]

Key points that emerged from the study included:

- Transdisciplinary approaches were essential, and broad contexts needed to be accommodated within the program's design.
- The latest teaching strategies were moving toward a more enterprising style, and enterprise education could be viewed in terms of "simply good teaching."
- Experiential and "curiosity-based" learning strategies were essential.
- Educators would need to be entrepreneurial and innovative in their own design and delivery strategies, so as to lead from a position of not only being knowledgeable but also being perceived as role models to enhance subliminal learning.[5]

In the broader context, and concurrent with the development of this teaching provision, the UK's Quality Assurance Agency for Higher

Education, whose "job is to uphold quality and standards in U.K. universities and colleges,"[6] was developing new guidance for the sector. This was announced in late February 2012 as "a new kind of learning," bringing together "experts in enterprise and entrepreneurship education to produce clear advice for institutions in supporting innovation and venture through higher education initiatives."[7] This quality standards group included a broad range of stakeholders and specialists, for example business school educators and art and design educators as well as representatives from government-recognized small business advisory bodies, student bodies, the national government, and enterprise research organizations. The broad stakeholder approach facilitated the trans-disciplinary thinking understood to be essential for development; it was not simply the domain of business educators and their associated perspectives. A range of reviews and reports from the international entrepreneurship research community were considered and discussed. One such example is the Enterprise Education Report for the Global Education Initiative for the Global Economic Forum. This suggested that "a greater emphasis is needed on experiential and action learning with a focus on critical thinking and problem solving. The pedagogy should be interactive, encouraging students to experiment and experience entrepreneurship."[8]

In terms of methodological stance, such contextual background had significant influence on the development of the national higher education quality-assurance guidance. Moreover, the advisory group was aware that the new guidelines would be available and suited to all subjects across the university sector. This was because learning contexts can be considerably different, and therefore if it was too inflexible, their guidance could be perceived by the higher-education community as positivistic and inflexible, and, as rules that dictate actions toward set outcomes, could be counter-productive. Subject contexts and styles of teaching had to be accommodated and considered alongside the overarching intent of helping make all higher-education students more entrepreneurial.

Specifically, the team was concerned that by attempting to generate new understandings that predict future opportunities, they may have created what legal specialists would call "rules of recognition,"[9] which are rules and regulations that restrict, rather than facilitate, innovation. Positivistic rules of recognition clearly do not sit well with intentions to deliver to a broad range of participants and to let them adapt and evolve their own approaches and meanings in their own contextual understandings and environments. Educators would need to develop their own ways of constructing knowledge, skills, and abilities, rather than simply following rules and procedures that might or might not fit their subject needs.

The authors of the new program and guidance took the view that consensus construction "that is more informed and sophisticated than any of the predecessor constructions"[10] was an appropriate stance, as there are many intervening variables,[11] and the diversity and heterogeneity of the sector[12] required integrating themes and enabling these through active engagement. The guidance also took into account evidence that management and leadership programs have been ineffective in developing creative-capacity and opportunity-recognition skills.[13]

The fact that an educator has to be a good interpreter becomes another factor to take into consideration. This assumes that the mental content is judgment-dependent, and creates what Byrne's philosophical discussion calls the "appropriately informed ideal interpreter."[14] If the educator is not aware or engaged in the creativity and innovation literature and debate, he or she will not only be unable to facilitate appropriate pedagogies but also might inadvertently provide inappropriate role modeling.

The perspectives that the new Welsh educator qualification and UK Quality Assurance Guidance bring to bear have been taken into account in new global initiatives. For example, United Nations Chief of Entrepreneurship Fiorina Mugione undertook a year of sabbatical study at University of Wales Trinity Saint David in order to refocus the organization's 'Empretec' study program in thirty-six countries.[15] The European Commission selected the educator qualification and its subsequent contributions to their Guide for Educators, "Acknowledging and Developing Entrepreneurial Practice in Teacher Training," as two of their thirty-nine best-practice case studies.[16]

Considering the Business School Perspective and Developing Creative Capacity

Creativity and innovation are often cited as important to enterprise and entrepreneurship, with many definitions explicitly showing the need for some kind of creative thinking. For example, descriptors from the literature and policy calls include:

- Learning to use the skills, knowledge, and personal attributes needed to apply creative ideas and innovations to practical situations. These include initiative, independence, creativity, problem-solving, and identifying and working on opportunities.[17]

- Embed in schools and higher education elements of entrepreneurial behavior (curiosity, creativity, autonomy, initiative, team spirit) already in primary-school education.[18]
- Now more than ever we need innovation, new solutions, creative approaches, and new ways of operating. We are in uncharted territory and need people in all sectors and at all ages who can 'think out of the box' to identify and pursue opportunities in new and paradigm-changing ways. Entrepreneurship is a process that results in creativity, innovation, and growth.[19]

Next, we consider that US-based business school environments have dominated the development of international programs,[20] with over 20,500 students taught at over 1,600 schools,[21] and that much of the foundation of enterprise education has been developed in business school environments.[22] Culturally, the business school approach to teaching and learning is significantly different from that of arts and humanities, for example in their assessment strategies, which are examination-led as opposed to project-led, and with delivery strategies that are frequently lecturer-led presentations to large classes, as opposed to developing incremental problems for students to solve through student-led experience.[23]

In 2007, the UK's Higher Education Academy's Subject Centre for Art and Design highlighted the fact that lessons learned from over one hundred years of creative pedagogical development are rarely borrowed by enterprise educators, and that pedagogies developed within the creative industries are rarely referenced or considered.[24] We postulate that due to this lack of engagement there is a considerable dearth of understanding of creative capacity development through divergent (widening) thinking strategies,[25] and that "most educational efforts emphasize convergent [narrowing] thinking, and therefore may do very little, if anything, for creative potentials."[26]

This argument of course is not new. Kirby[27] posits that the traditional business school education approach "stultifies rather than develops the requisite attributes and skills to produce entrepreneurs." Kirby is among a growing number of senior enterprise and entrepreneurship educators who advocate the development of right-brain (creative) entrepreneurial capabilities, in addition to left-brain analytical skills, to encourage and stimulate the entrepreneurial imagination.[28] Recent texts support the premise that entrepreneurship is both a science and an art[29] and that assessment strategies need to reflect this split perspective, a point we will expand on below.

Integrating Design Thinking from the Outset

With the above in mind, this essay adopts what is becoming understood as a "design-led" approach, which is "essential in problem-solving scenarios and, therefore, relevant for business managers."[30] The stance of the design thinker is well established, as is design education, however in recent years business schools have also started to embrace the process, terming it 'design thinking'. Although various models exist, the term was popularized in 2009 by the Dean of the Rotman Business School in Toronto,[31] who claimed it to be the next competitive advantage; however, we must not initially confuse design thinking with design-led activities, because this might lead the reader to think them to be one and the same. As Cunningham notes, the World Economic Forum's Global Competitiveness Report 2001–2002 highlighted that "there is a distinct correlation between design-intensity in enterprise activity and product development and broad economic competitiveness."[32] By this we see that design-led business activities are considered to be advantageous, and that these types of activity can impact on competitiveness. This highlights some potential questions, such as what may have been done differently in design education that could lead to this type of impact, and could adopting the pedagogical of a design educator have benefits?

To take this a step further, in addition to Martin, the term "design thinking" is sometimes based on principles established at the IDEO design company in Silicon Valley, which was conceptualized and developed by industrial designer Tim Brown, who took traditional design approaches and applied them to business development so that the rigor of the process could be better considered. His texts took what was being done in the creative industries and translated them into process models that could be interpreted and employed to solve business problems, featuring what Stanford University describes as design sensibilities and methods that match consumer needs and convert ideas into market opportunities and viable businesses.[33]

We agree that "entrepreneurship as a discipline is fragmented among specialists who make little use of each other's work"[34] and, moreover, that "each discipline views entrepreneurship from its own perspective without taking cognizance of approaches in other disciplines."[35] Research is only one aspect of education, however, and evidence suggests that professors rarely concern themselves with different ways of teaching.[36] Therefore, presenting in a "design-led" style that is multi-faceted and interdisciplinary could help unravel some of the puzzles facing enterprise and entrepreneurship educators, especially if it actively demonstrates the approach in action.

We contend that design education has much in common with early-stage enterprise education, as "the essence of design education is to establish their [students'] reservoirs of experience, fostering creative thinking processes for originality and novelty."[37] We see Harvard Business School's Interdisciplinary Mind, Brain, Behavior Initiative as a support for this design-based stance. Working on the premise that artificial categorizations between disciplines "do not reflect how people actually live their lives," their initiative postulates that, "the most promising knowledge frontiers typically exist at the boundaries between fields rather than at the fields' respective cores."[38]

The practitioners of design thinking, and educators within the design field, are cognizant of the fact that creative outputs are dependent on the development of divergent thinking strategies and ways of assisting enlightenment through the production of as many alternative solutions as possible.[39] This enhances the capacity to make new connections and associations,[40] which in turn develops skills in opportunity recognition and innovation. When students are encouraged to find singular or "correct" answers, such as those found in examinations, the opportunity to develop creative alternatives is limited. Moreover, if the aim is to accommodate recent trends and contextual decision-making within solutions, such fixedness makes the examination as a type of student assessment irrelevant. As things can and do change, sometimes rapidly, the learner's interpretation of the moment within which he or she is being assessed and evaluated needs to be an integral part of the evaluation. It is less a question of the "right" answer, and more one of how many alternative answers a student can perceive, and how well-aligned this range of solutions is to the identified problem or issue that they wish to address. Also, due to a rush to arrive at quick solutions, "premature articulation" occurs, a situation where potentially useful alternatives and perspectives have not been adequately considered.

In presentations, we have used examples of divergent connections that occur in a learner's mind to illustrate this type of thinking through direct experience, so that those in attendance can experience the phenomenon, before coming to understand that it is supported by theoretical constructs. In order to emphasize the breadth of neurological connections being made and processed in the minds of those who attend, we facilitate emotional responses that enhance memory and learning. Importantly, we mirror our experiences of design education and do not explain the theory first, but let it make sense once those in attendance have experienced the issues for themselves. In short, those who attend rely not only on the new information being shared with them but also on prior personal experiences that preceded the dissemination of the relevant knowledge. This approach is designed to cement the learners' understanding through what is known in educational psychology as "associative recall."[41]

To explain and demonstrate this approach, we will use the example of a presentation made to delegates at the 2012 AUC conference, "Entrepreneurship and Innovation: Shaping the Future of Egypt."[42] Our keynote presentation opened by contextualizing and visibly linking "up-to-the-minute" speaker and delegate comments from the previous day's presentations, so comments and contexts could be considered that were only hours old, not drawn from textbooks written sometime in the past. The audience commented on leadership and visioning opportunities, the importance of innovation and independent thinking, and the need for teaching approaches that develop opportunity-recognition skills. From the outset, the presentation addressed these emerging conference issues and was adapted accordingly.

We took this approach because it is widely recognized that entrepreneurs lead by example, are not risk averse, and are able to respond to the unexpected by drawing on whatever resources and knowledge they have at hand. Hence this approach illustrated the process in action and demonstrated through doing, not merely telling. This is important because students learn as much from what educators do as role models as from what they say.

The next section of this particular presentation was designed to demonstrate the process further. Delegates normally anticipate questions at the end of a presentation, so they were invited to ask questions from the outset to help to shape the dialogue. This was acknowledged in feedback to be a better way to engage the delegates, and actively demonstrated the presenter's propensity for managing risk. This presentation strategy also indicated how role modeling could be accommodated within a learning scenario, because delegates would become aware of their own learning subliminally, through experience as well as formal recollection. As questions from the floor preceded the full presentation, there was an understanding that the dissemination strategy was not merely sharing information but responding to specific inquiries that the attendees had become curious about. Thus, curiosity and the immediacy of the moment were key elements of the presentation. If attendees of an entrepreneurial education session do not develop a curiosity about the topic, or a wish to develop deeper understanding, the presentation can be no more than a "telling about" as opposed to a "preparing them for" entrepreneurial action.

Creativity and Context: Clarifying the Demand for Interdisciplinary Thought

To further illustrate our concerns that there was a lack of interdisciplinary understanding between design and business education into our discussion at AUC, we drew upon Cooper's long-established "Stage Gate" theories and Bessant and Tidd's investigations and theoretical stances on innovation in entrepreneurship.[43] These only indicate the importance of ideas in the introductory stages of a business, and neglect to highlight that whenever an unexpected problem arises, it is the ideas generated to solve the problem that take the business forward— hence, creativity is essential throughout the process unless the business is long established, stable, and unlikely to have to respond to change. To illustrate this, we populated a typical stage-gate graphic with indicators showing where creative and innovative thought was needed (fig. 3.1).

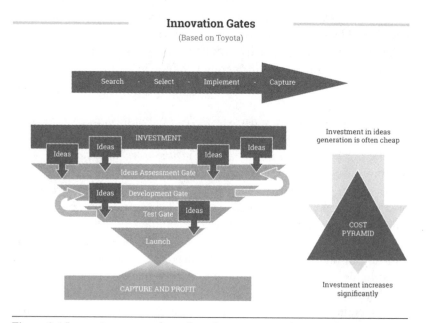

Figure 3.1 Innovation gates and cost factoring

This graphic further illustrated the fact that creative thinking, far from being important only at the outset, permeates all levels of business generation. Moreover, the more conceptual thinking is undertaken at early-stage development, the less of a negative impact there is on the budget. If multiple alternative concepts are considered earlier on, there is more room for change and adaption in response to pre-identified potential problems, thus saving considerable costs in development and production.

During the AUC presentation, we also developed the premise that intrinsic motivation is a key feature of the entrepreneurial mind. It is the internalization of problems to be solved and the relevance of them that drives an entrepreneur, not the extrinsic motivation of reward for doing well.[44] Both entrepreneurs and learners have to be tolerant of ambiguity, willing to persevere, motivated to grow, and ultimately to work and plan with the courage of their convictions. While readers will readily align these characteristics with the literature on enterprise and entrepreneurship, we draw them rather from Sternberg and Lubert's discussions on teaching that develops creativity.[45]

There are many studies on the development of creative capacity and alternative metrics for evaluating enhancement. Perhaps the best-known is the Torrance Test of Creative Thinking (TTCT), which was "renormed" to reflect cultural changes five times between 1966 and 2008. It is important at this juncture to recognize that creativity is distinct from intelligence, and that while international IQ tests have increased in rankings in the last century,[46] there are concerns that educational constraints such as normalization through Scholastic Aptitude Tests (SATs) in high schools have resulted in a marked decrease in creative-thinking capacities. For example, in a detailed discussion about using TTCT tools, Kim describes a "steady and persistent decline" in the creative capacity of pupils in US schools between 1990 and 2008.[47]

Putting this in an enterprising context, Pink describes what he considers to be the three stages of the industrial age.[48] He details the initial, historic move from a need for physically strong workers toward those with knowledge and understanding, drawing upon the well-used phrase, "the knowledge-based economy." He contends that with the proliferation of the Internet, such thinking is out of date, as knowledge on almost any topic is now freely available via chat rooms and websites. He goes on to suggest that this presents a new problem: How can we make sense of such a confusing array of wide-ranging and easily available knowledge? Pink concludes that in many ways, we are now in an environment where business survival is less about retaining knowledge and more about enhancing knowledge-harvesting capacities that can be adapted to specific contexts and situations.[49] Recognizing patterns in situations of complexity is a key aspect of creative capacity, and as will be shown in our discussion on curiosity-based learning, this features strongly in assessment strategies within arts-based education.

To illustrate our points further at the conference, we presented a critique of Schön's and Kolb's reflective thinking,[50] which brings to mind the limitations of differing value mechanisms and mental computations, especially when we consider what we believe to be concrete experience. Reflection often enables us to see things differently, and to see how what may at first appear to be concrete knowledge can be actually be an assumption. Reflection can also lead us to understand what we may have thought or done subconsciously but can only bring to mind after the event and through closer investigation. New knowledge can assist and compliment existing understanding, but it may also undermine it when it conflicts with prior assumption.

For example, the concept of subliminal knowledge harvesting described by Pink introduces things that we did not realize we knew, things that are frequently linked to our subconscious strategies of learning but rarely surface until supportively reflected upon.[51] For example, in deductive reasoning, testing in new situations requires all facts to be in place. However, in real world terms, and when considering computational coherence for future planning, many facts may be unavailable. Future thinking suggests that best guesses may have to suffice, which are often referred to as hunches or gut instincts. In practical terms, and within the paradigm of abductive reasoning, we can model future scenarios by knowingly and consciously utilizing assumptions when hard facts are unavailable, which in turn lead to an enhancement of future visioning skill development.[52] Here all kinds of assumptions, as long as they are clearly expressed, replace known facts and help the thinker foresee multiple opportunities and alternative solutions.[53] Thus, in some cases what is perceived as deduction is more correctly defined as abduction, because all facts cannot be not known until the moment has passed.

This discussion required some expanding upon at AUC. As subliminal learning is crucial for our arguments, our primary aim at this juncture was to illustrate this phenomenon through direct experience as opposed to simply being told about it, because we were once again using the design educator's paradigm of learning through practical experience and not just theoretical constructs. Delegates were invited to look at a set of Russian derivations of the Cadbury brand and its associated logos. Despite the fact that the name in Russian is significantly different visually, as it starts with a K not a C, and does not read as Cadbury at all, a high proportion of delegates indicated that they had recognized the brand immediately. This introduced the problem that, while they had such knowledge, they were unaware of how they had learned it. Despite high levels of recognition, no delegate reported that they were able to rationalize his or her own thought processes until they had carefully reflected upon how they had made the seemingly obvious connection through complex associations of color, shape, and flow.

A secondary aim of this exercise was to arouse curiosity and set the scene for further discourse. Termed "curiosity-based learning," this strategy extends the pedagogy of problem-based learning[54] by enabling learners to recognize new problems for themselves, much like an inquisitive entrepreneur. Through this exercise, the delegates became aware of their own limited understanding of how they make decisions such as recognizing the Cadbury brand. One benefit of the technique is that it enhances intrinsic motivation. The learners wish to know more in order to solve a problem of which they have taken personal ownership. New knowledge is assembled as the learner/delegate participates in what they might initially consider to be an unstructured experience, one that only makes sense some time later, after their subconscious thinking has been explained and drawn out. As Azer contests in the field of medical education, highly structured educational experiences may suit tasks that can easily be broken down into sequential stages, but they do not help develop appropriate responses to highly intellectual decision-making processes that may go beyond conscious thought.[55]

Following the Cadbury logo exercise, to further demonstrate how diverse creative connectivity within the brain often occurs at subconscious levels, the authors introduced the concept of "bisociation," the combining of two new and/or alternative conceptual understandings. Our strategy was to lead to another experience of an 'aha' or eureka moment in the learner's mind. In this example, the slide showed an advertising slogan that said "Kills Bugs Fast." In the sense-making interaction with delegates that followed, we drew out a wide range of responses that indicated potential products or services that such a text might promote. Most thoughts centered on pest and insect control or computer viruses, but none guessed the actual product being promoted. As an image of a Porsche sports car slowly appeared on the screen, they each made the connection to flies on the vehicle; the puzzle was solved in their own minds and the smiles grew in the room as the unexpected answer came to them. It was not a wrong answer but rather one that had not surfaced through rational thought. Moreover, they had not been told that this mental connectivity worked. They experienced it for themselves and rationalized it through reflection. This power of unconscious thought is often overlooked, yet it can yield more creative solutions as it is more associative and divergent.[56]

Hence, we illustrated that we often see the whole but don't stop to work out how we have managed to do so, let alone challenge our underlying perceptions. Developed by German psychologists around 1910,[57] gestalt theory helps illuminate what we learn subconsciously though associative mental procedures, and is a common component of many art-based studies. Michotte is quoted as suggesting that "the contents of thought are nothing more than the associations and re-combinations of sensations and images."[58]

The example of adding value to watermelons was offered to AUC delegates to consider. Growing the fruit in square boxes (with a resulting square shape) saves them from damage (as had been witnessed by delegates seeing a truckload fall off on their way to the conference), and enhances their perceived and unique value through their capacity to be stored in larger volumes safely, even without any packaging. Another example showed pyramid-shaped watermelons, which were sold to tourists at significantly higher prices. In both cases, seeing beyond the obvious brought about unique selling points that offered new value to an otherwise common produce. Disparate connections such as being able to see traditionally round melons and their traditionally square packaging systems as an opportunity as opposed to a problem challenged underlying assumptions.

Creativity, Evaluation, and Assessment

It is recognized that assessment methods and associated value metrics, when applied to enterprise and entrepreneurship courses, do not always match,[59] especially where creativity, innovation, and opportunity recognition are required.[60] Process-oriented and solutions-focused learning philosophies[61] can alternatively be viewed as the art or the science of enterprise,[62] which helps us rationalize the disparate nature of the two approaches. Recent research indicates that solutions-focused assessment has been the main vehicle of evaluation, and that there is a significant need for more innovative assessment strategies.[63] Moreover, assessment practice makes little distinction between significantly different learning approaches,[64] and so may not be constructively aligned in terms of the task matching the evaluation.[65]

If exploiting creativity, intelligence, and knowledge are key features of the entrepreneurial mindset, and creativity requires solutions that are new and innovative, then solutions-focused assessment would require the educator to have a predetermined solution against which to measure students. In such a case, the educator would be creating the new solution from which to gauge the students' success, rather than gauging success against a student response or solution that the educator had not foreseen. Clearly, then, this assessment strategy cannot work.

If discovering unusual, unexpected, and novel solutions to problems is our goal, we have to review these metrics. Developing the ability to question the validity of an initial problem leads the learner to discover new problems. These can either be a subset of the original problem or result from an in-depth understanding that an incomplete or poor observation underlies the problem.

To illustrate this, we can consider the issue of a hospital asking how it can afford more cleaners to ensure healthier, cleaner environments to counter increasing levels of infection. Abductive reasoning tells us that there are assumptions in the question. For example, does everywhere have to be cleaned to the same level to ensure that infections are not spread? This underlying thought leads to a new question: Are there specific environments or situations that carry the infections? If the answer is yes, then perhaps frequently handled items, such as door handles or television remotes, could come under investigation and a more focused, perhaps cost-effective solution could be derived? Developing the capacity to elicit new and potentially unseen solutions, as no one else has spotted the problem from which they are derived, offers opportunity for a higher level of learning. In this example, evaluating outcomes against perceived notions of correctness becomes extremely problematic, because new insights have emerged by questioning the question. An educator may be asked to consider how they can evaluate a response that invalidates the original question being asked.

Of course, evaluating enhanced creative capacity development is not new, so delegates at AUC were invited to consider an assessment practice known as "divergent production." Loosely based on Torrance[66] and psychology-driven literature,[67] pragmatic assessment strategy contains three key components. First, and perhaps best recognized, is "ideational fluency," where learners are challenged to come up with as many alternatives to a problem as possible. Typically represented by exercises such as "How many ideas can you come up with for a paper clip," the intention is to prepare the mind for multiple and diverse solutions. As will be discussed, this is often a short-term and "once in a while" exercise, and is rarely built upon and repeated to enhance capacity. The second component is "expressional fluency," a form of reflection that requires a grid or framework to be drawn up that shows the range and variance of connections that formed in the mind. Conscious and, where possible, subconscious connections are drawn out to clarify the triggers and connections that led to the range of ideas expressed. Single solutions are not acceptable, and multiple solutions grow in number as the learner enhances his or her abilities. Much like a good chess player, learners are expected to plan ahead and second-guess potential issues, thus delivering a range of solutions that can match different scenarios. The use of an expressional fluency chart also helps avoid the scenario of multiple forms of a single idea. The third component, "divergent production," encompasses all of the above, as the educator or evaluator can now consider the breadth and diversity of the range of solutions.

We summarize this below, using the example of a watermelon to illustrate how this can be accomplished:

- Ideational fluency: Evaluate and progressively increase the complexity of a series of questions that elicit multiple responses; for example, how many solutions can a student think of to a given problem? (Such as, how many uses can a student think of for a watermelon?)
- Expressional fluency: Within the proposed uses above, how many overt connections can a student make that justify the decision to use the example? (This assessment provides for both deep (focused) and wider thinking (for example, from another subject discipline or perhaps human-centered emotional understandings), offering proof of the connecting thought processes that led, for example, from thinking round (watermelon) to thinking square, and seeing the change of shape as an opportunity, not a problem.)
- Divergent production: How broad and diverse can the students' mental connections be? (Ideational and expressional fluency combine to prove disparity and breadth of thought. The more similarity between the connecting thought processes expressed, the less creative and innovative they are likely to be; thus, the extension of thinking from round melons to square, and then on to pyramid shape, illustrates the point that wider thinking has taken place.)

Thinking about Thinking Is Not New

Most of our discussion here has been drawn from the literature on creativity and novelty. It is not new, but it may be novel to the enterprise educator. For example, computational and strategic thinking, or rather thinking about thinking, was discussed in some depth by the American Psychological Association's Task Force on Psychology in Education in readiness for school redesign and reform in 1993, but it was never fully implemented.[68] Robert Sternberg and Wendy Williams's approaches to teaching creativity discuss the questioning of assumptions; defining and redefining problems; tolerating ambiguity; and identifying mistakes from which to learn.[69] Torrance discussed the creative environments needed to enhance idea generation.[70] Merle Karnes voiced the belief that creativity could be developed through instruction,[71] and Jami Shah et al., discussed metrics for measuring the ideation process that encompassed novelty, variety, quantity, and quality.[72] Preceding all of the above, the Franklin Institute in Philadelphia utilizes an awards system that was originally set out in 1824 by a committee for inventions, awarding prizes for uncommon insight, skill, or creativity.[73]

Creativity, Cognition, and Deeper Thinking

Another example of silos of understanding and expertise that have not permeated entrepreneurship and business education are those concerning cognitive science. Cognitive-affective models have been in existence for some time, but significant advances in brain scanning techniques have enabled us to gain deeper insights into brain functionality and to test theoretical constructs. Creativity evaluation in relation to neural processes, and an enlightened understanding of how occupational therapists with specialist skills deal with a patient's brain injury, add novel insights.[74] We know from expert neuroscientists such as John Kounios of Drexel University and Mark Jung-Beeman of the Cognitive Brain Mapping Group at Northwestern University that brain cell death is not irreversible, and that the plasticity of the brain means it is constantly changing and reconstructing itself to cope with experiences and ever-changing demands. This lifelong self-adjustment and self-optimization process continuously remodels the cognitive, behavioral, and emotional status of the brain.

These insightful approaches lead to 'aha' or eureka moments that "contradict the classic model of learning in which the learning process was assumed to be gradual."[75] Kounios has demonstrated that the conscious brain only reacts to the subconscious discovery of a solution. This moment is known as 'aha,' and it only arrives when we are emotionally prepared, usually in a relaxed state of mind.[76] Thus, understanding the phenomenon of insight enables us to develop discovery strategies within learning environments.[77]

Emotion can play a greater role in both decision-making processes and idea generation than is normally understood. As noted by Zaltman, "studies of the effects of brain lesions demonstrate that when neurological structures responsible for either emotion or reasoning sustain damage, the affected individuals lose their ability to make the kind of sound decisions that permit a normal life."[78] Moreover, scientists in the USC Viterbi School of Engineering report that they can switch memory on and off at the base of the hippocampus in rats' brains by embedding a micro-switch. As a rat's hippocampus closely matches the human limbic system—the part of the brain that controls our emotions—it offers new insights into the role of emotions in learning, memory, and behavior.[79] In essence, detaching emotionally relevant brain processing from what appear to be logical decision-making processes results in diminished reasoning. The reader may also wish to consider how the fact that emotional responses help cement memory sits with the mind of the entrepreneur in terms of learning strategies. As emotional responses help cement memory, this has implications for entrepreneurial learning strategies. Behavior, too, is influenced by emotion; as anyone who has recently been involved in a heated argument will tell you, in later moments of calm they will come up with far better arguments than those used at the time of the confrontation.

Martinez et al., inform us that little is known about what elements of training and studies are most valued by entrepreneurs themselves,[80] though there is general acceptance that transmission teaching in lecture environments is ineffective. Hence, experiential approaches are the preferred method in classes where developing entrepreneurial capacity is a goal.[81]

In arguments not too dissimilar from those expressed earlier in this text, classical conditioning theory argues that the most basic type of learning is implicit, and we may not be aware of all of the associations we are making to reach conclusions.[82] The Pavlov experiments where dogs salivated upon hearing a bell ring illustrate this subconscious decision-making process.[83] Subsequently, educators may not be aware of actions they make that impact learning nor of the differing impacts of different types of learning. "Procedural learning," for example, enables us to learn to ride a bike without really analyzing what we are doing but is distinct from semantic and episodic memory, where recollection is forced through repetitive tasks or linked to episodes that help recollection. This has been proven through research with amnesic patients who cannot retain short-term memory yet can develop new skills despite not being able to recall the lesson or class.[84]

To take this discussion a step further and to clarify its relevance, a part of the hippocampus—one of the earliest evolutionary parts of the brain and widely accepted to be associated with emotion—is good at acquiring arbitrary and relational links in the mind. This kind of brain wiring is known to assist memory tasks that "tend to be non-episodic and can be mediated by arbitrary and conjunctive operations."[85] Known as "cued recall," this helps bring disparate things to mind, especially within novelty detection scenarios. Our brain works differently when we recall knowledge than when we bring to mind disparate but useful connections. Hence, experiential educational episodes have a key role to play in more ways than one; the educator's role modeling teaches as much as his or her delivery of educational facts and knowledge.

We thus argue that subconscious mental activity impacts our decision-making processes, and if we want our students to behave and respond entrepreneurially, we need to act accordingly in our learning environments. Bear in mind that students are often more interested in external role models, such as real entrepreneurs, arguably because they are experiencing strong and emotionally driven new episodes of memory, not because they are forcing themselves to recollect semantically.

In terms of behavior change, which neurons the brain decides to discard through inactivity is as important as those it develops, because "with every new experience your brain slightly rewires its physical structure."[86] This also posits a view that a good role model can inspire his or her learners, while a poor role model can work against them. Again, this is far from new; as Renzulli contends, the role-model teacher is able to reward unorthodoxy, originality, and creative conceptualization, because he or she has the interest of his or her students.[87]

Evidence also suggests that there are strong indications that "an individual's brain hemispheric processing mode, known as hemisphericity, is directly related to that individual's learning style. Therefore, hemispheric specialization and the resultant learning style have significant implications for learning and teaching."[88] If learning experiences are challenging and sit within a social context, they are highly compatible with brain-related learning environments, as both sides of the brain are employed.

To demonstrate this concept, delegates at AUC were shown images and diagrams of physical differences between the micro-molecular structure of the right and left hemispheres. The left side, often considered linear and solutions-focused, has a characteristic that resembles, metaphorically speaking, the narrow branches of fir trees (fig. 3.2), while the right has more in common with oaks. The soma (cores) of these neurons are connected by long branches known as dentrites, or dentritic strands. The key point is their differing footprints and their ability to reach and connect to other neurons, as their potential for connectivity is much broader in the right brain than in the left.

Experts at Midwestern and Drexel universities in the United States have also been undertaking a series of experiments to uncover new perspectives on language development and what they describe as the "prepared mind."[89] Focusing on the 'aha' moment when a new idea consolidates itself in the mind, they have been able to track how this occurs and to articulate theories that support many previous assumptions. Understanding these physical differences has been a key discovery, enabling them to postulate ways in which connectivity between neurons can lead to broader and more creative thought processes.

At the base of the brain, and most probably one of its most early evolutionary developments (as it is most closely linked to the spine and therefore one of the first receivers of sensory data from the body), is the amygdala. "One of the functions of the amygdala is to interrupt ongoing activity in order to induce quick responses to dangerous situations."[90] This "priority call" takes precedence over other cognitive functions, and also relates to the physical structure of the brain (fig. 3.3), because it interfaces directly with the hippocampus—the part of the brain that manages memory. In simple terms, this could be described as the emotional thermometer of the brain, as it determines what information is stored and what is not.

In practical terms, this connectivity can perhaps be understood best when we consider classroom situations that potentially limit learning through the creation of situations that stress the student. As one of the major features of enterprise education is to develop learning "in situations of ambiguity and risk,"[91] this aspect has to be understood. Pitching and presentations are good examples of this kind of situation, and if managed well by the educator as frequent and part of a developmental approach, can result in a less stressed learning situation that is initially non-confrontational but builds over time to enhance resilience and a capacity to think on one's feet. This requires mentoring that facilitates behavior change; it is not akin to learning facts but more like training and coaching an athlete.

Key Point - Developing Synapse Potential Through Learning

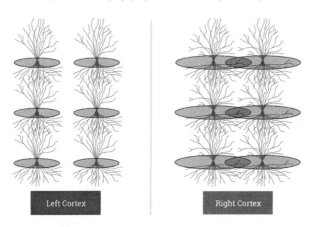

Micro molecular structure images courtesy of Mark Jung Beeman and
John Kounios, Mid Western and Drexel Universities

Figure 3.2 Comparative details of the molecular structure of the left and right sides of the brain (images reproduced by kind permission of M. Jung Beeman and J. Kounios)

Key Features

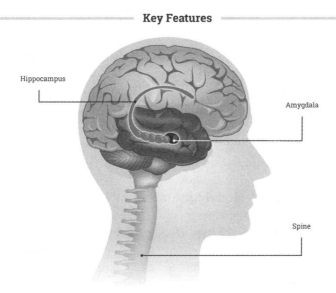

Figure 3.3 The physical relationship between the spine, the amygdala, and the hippocampus (the limbic system)

This "preparedness of the mind" research also triangulates well with evidence from other methodological sources. For example, Fasko[92] cites earlier research from Parnes and Noller,[93] in which a longitudinal study of freshmen undertaking credit-earning studies in creativity consistently and significantly outperformed a similar-sized control group of 150 students when evaluated beyond the traditional exam scenario. Among other traits, divergent thinking, flexibility, problem solving, problem prevention, and responses to problems in realistic scenarios were highlighted as significant areas of improvement. Perhaps surprisingly, there was also a correlation with success in other studies compared to the control group. The enhanced ability to evaluate ideas was another significant finding, once again indicating that starting by evaluating ideas for a business plan could be significantly flawed.

Amabile et al., point out that there is a belief in business that working under pressure helps consolidate creative discovery. They show that this is far from the truth, however, suggesting that "when creativity is under the gun it usually gets killed."[94] Moreover, they say that cognitive paralysis sets in when creative solutions are needed in situations that are highly time-sensitive, or when constant interruptions break the chain of thought. Importantly, there can be a hangover effect for days, as the brain has to readjust itself to a more balanced thought process where it is not pushed by the rush to a conclusion but given the time and opportunities to think far and wide again, which is the process of creativity. Hence, emotion and the limbic system are once again seen to impact on learning.

To conclude our discussion on cognition and deeper thinking, we note that the UK's Independent Practitioners in Advertising describe learning in ways that acknowledge hemispheric functionality as "diagonal thinking."[95] As no relevant tests were available from the university sector or other stakeholders, and as this is what they wished to test, they undertook their own research and came up with their own model of assessment.[96]

Design Education: A Model for Enterprise Education?

If we combine the above perspectives and set them into a design-thinking context, a picture emerges. Typically, design education (as shown in our discussion of assessment through divergent production) requires multiple decision-making processes that should offer multiple solutions that may be a fit for differing scenarios at any given point in time. This is in order for the thinker to be flexible and adaptable. Figure 3.4 shows a typical approach to design education.

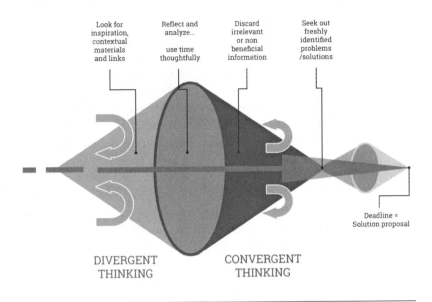

Look for inspiration, contextual materials and links

Reflect and analyze... use time thoughtfully

Discard irrelevant or non beneficial information

Seek out freshly identified problems /solutions

Deadline = Solution proposal

DIVERGENT
THINKING

CONVERGENT
THINKING

Figure 3.4 Divergent and convergent thinking as located in design education strategies

As indicated by the opening cone, initial stages are designed to let the mind wander and seek out inspirational sources. This is usually during the early mission stages, when time is less pressured. Known as divergent thinking, this period concerns itself with the breadth of idea generation and is deliberately non-judgmental. Students are encouraged to extend this period as much as possible, and to avoid "premature articulation," which refers to going for an early idea (usually less thought through) before considering more diverse opportunities. Following a period of reflection, ideally at a time when students are engaged in other activities and less stressed, the convergent ideas evaluation commences. During this time, and if the learning situation has been well managed by the educator, a shortfall in knowledge or missing information should be self-discovered by the student. This is the 'aha' moment of learning, when self-realization leads to new problems and issues. To solve them, the procedure repeats itself and the student returns to an exploratory mode of thinking. This may happen many times before the final deadline is reached.

Importantly, Kounios considers that learning to develop insightful strategies is not at all like the step-by-step approach that is commonly employed in education, as it will "contradict the classic model of learning in which the learning process was assumed to be gradual." [97] Kounios' positron emission tomography (PET) imaging scan experiments show that the conscious brain only reacts (in the right anterior superior temporal gyrus) to the subconscious discovery of a solution formed elsewhere within broader neurological connections. Logic does not create it, so assessment metrics that use traditional logic may fail. Insightful, non-algorithmic approaches to studies and assessment are required.

This process can be illustrated through a quick and simple example of a typical design student's task— coming up with a new logo or brand design. This will require considerable research and understanding of the target audience and the aims of a company, which entails divergent thinking and associative breadth of thinking (gestalt). Following the production of a number of raw prototypes, often termed roughs or scamps, the designer will attempt to rationalize his or her thinking and focus on solutions, the typical convergent thinking scenario. At some point thereafter, the designer discovers that the logo will feature on the Internet and television, offering him or her the opportunity to animate the design. Now the designer has to revisit the idea in a new way, breaking down the components and developing new concepts, and divergent thinking skills once again take priority. Once multiple prototypes are considered, they are checked against technology to see which are workable. Convergent thinking thus eliminates weaker solutions. Typically, clients are offered multiple alternatives, justified by the designer's reflection on his or her own work and influences (divergent production assessment). Client input normally requires further work, and so the story continues.

The process aligns well with the neurological and other considerations voiced earlier. As can be seen in figure 3.5, the large ellipses at the center of each stage should capture multiple ideas, then, through evaluation, lead to the discovery of shortfalls. These points can be considered as prototypes, when new learning will enhance discovery. Our blue arrows indicate the points at which the molecular structure of the brain aligns with the different nature of the tasks, when breadth instead of depth is the preferred outcome, and connective 'aha' moments lead to new discoveries by the students themselves. [98]

Now let us align this to the world of entrepreneurs. They have ideas that challenge norms and open doors to opportunity (divergent thinking). They ponder these and attempt to eliminate potential issues from their range of

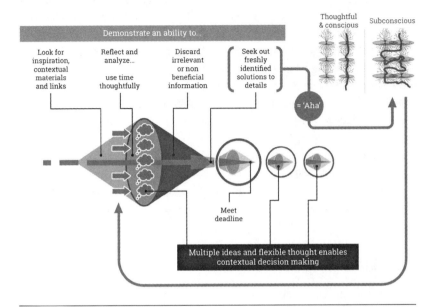

Figure 3.5 Convergent and divergent thinking in the context of neurological understanding

ideas (convergent thinking). Along the way, new knowledge or changing circumstances invalidate some of their ideas, so they start to review them in a new light—another example of divergent thinking. Having found solutions, they revisit the marketplace and create a business plan (convergent thinking). The plan is piloted and tested through initial trading (divergent thinking led by discovery), and then the results of their prototype business are tested against their plan and reviewed accordingly (convergent thinking in response to new situations, knowledge, and experience).

This model ties in well with the recent UK government-commissioned Review of Business–University Collaboration, which states that "enterprise skills require responsiveness to unexpected pressures and tasks; they require reaction to changing circumstances and disruptive interventions. These attributes are contrary to the established framework of assessment processes. Enterprise skills do not presently lend themselves to formal assessment methods."[99]

Unlike many counterparts, design educators are used to not offering full scheduling details to their learners, and their approaches may appear unstructured. As can be seen in the figures, however, this is far from the case. The model permits unexpected challenges, such as additional tasks. These become apparent through the presentation of new contextualized and relevant changes, new information, and moments of 'aha.' Far from being bad planning, this is deliberate, as it mirrors the life experience of the designer. To facilitate this, the UK Quality Assurance Agency Subject Benchmark Statement that guides all higher education providers of design education includes the following:

2.6 Experiential, activity and enquiry-based learning are features of the art and design curriculum in HE. Through this approach, students have been encouraged to develop both the capacity for independent learning and the ability to work with others. Students not only develop the ability to solve set problems in a creative way, but they also develop the ability to identify and redefine problems, and to raise and address appropriate issues.

3.10 Design is an activity of creative reasoning that is dependent upon flexibility of ideas and methodologies informed by an awareness of current critical debates. It ranges between the expressive and the functional... is also an iterative process based upon evaluation and modification. Design is reliant upon constantly evolving dialogue and negotiation between the designer (working individually or within teams as proactive collaborator/mediator) and the client, manufacturer, audience, user, customer, participant, or recipient.[100]

Thus we make our contention: If we are seeking to constructively align our assessment to the learning tasks we create,[101] design education experience and guidance has much to offer to the enterprise and entrepreneurship educator. While these arguments can be extended and developed to encompass issues of empirical work and evidence of design's success in the marketplace,[102] the aim of the AUC presentation was to enlighten the delegates and enhance understanding through practical experience as well as theory construction. Hence, most of the cognitive strategies discussed here were actually employed during the conversations.

For example, to help close our argument that design's "constructed reality" approach is valid,[103] we illustrated the early success yet forthcoming demise of the Pringles snack chips can, which, due to the changing perception of its value from a secure airtight package to something that has become difficult to recycle within sustainable energy strategies, triggered the question, "Why do we always do it that way?" We also showed through incremental design stages that the space shuttle, due to the size of its rocket motors and transportation restrictions, was effectively based on two ancient horses' rears: The US rail systems that transport the motors are based on UK systems of 4 feet by 8.5 inches wide, which in turn are based on horse and cart technology. These designs and technologies emanate from the chariots of the Romans, Greeks, and Egyptians, where the width of two horses' rears in harness led to an optimum width for their chariot wheels, a restriction that still limits our reach to the stars.

A concluding slide (fig. 3.6) discussed the centrality of developing an enterprising mindset.[104] Our diagram of the "Enterprising Angel" illustrates our perspective that intrinsic motivation is key to innovation and

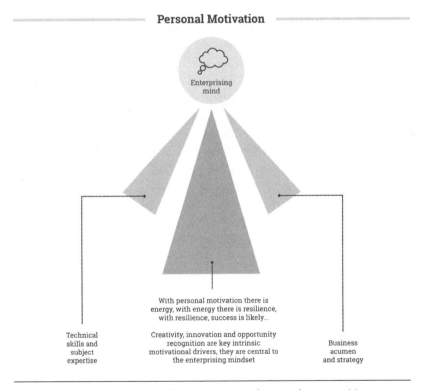

Figure 3.6 The enterprising angel perspective on educational opportunities

creativity,[105] and that perseverance and resilience are linked to the motivation to succeed. This central core, supported by technical and subject-specific knowledge and business acumen, is the model from which our entire discussion has evolved. The authors believe that enterprising people and entrepreneurs can be encouraged, developed, and enhanced though their own motivational development, led by educators who raise curiosity and empower their students through augmented understandings of creativity and the way it works in the mind.

Closing Discussion: Approaches to the AUC Presentation

Using a design-led approach, our discussion and delegate interactions aimed to inform and enlighten through personal experiences as well as through the more traditional presentation of facts. Our text aims to assist the reader to understand the presentation style we employed, and to indicate some of the approaches already used extensively in design education. We have discussed the depth of available research yet indicated that these understandings are rarely referenced in business-teaching scenarios and enterprise-related literature. In keeping with our design-based approach, we used visual metaphors and decoded metonyms using watermelons, Pringles canisters, and horses' rears. Some of the presentation contained humor, as this helps the disposition of delegates by shortening the power-distance relationship between presenter and listener.[106] Frequently, the delegates actively contributed to the discussion, and this introduced occasions to manage opportunity, ambiguity, and risk during the presentation.

Our presentation hoped to illustrate the informed and complex practice that breaks boundaries and actively takes into account important aspects of consensus within a wide range of subject contexts. As opportunity recognition is recognized to be a key skill of the entrepreneur, and we believe there is a dearth of understanding of creative capacity—especially in terms of neurological understandings and other prior work—much of our discussion focuses on what we consider to be missing or thinly evidenced discussions. Emotional constructs and a practical understanding of the functionality of the brain were key issues of contention.

In terms of role modeling, and delivering the creative breadth as opposed to depth of learning that is needed to be entrepreneurial, we turned to principles of constructive alignment[107] and "fit for purpose" assessment strategies. We also offered specific and relevant examples of where and how the bringing together of diverse silos of specialist expertise led to new learning, from design education to business education, and from occupational therapy to brain mapping. All these have a part to play, yet they are rarely brought together in a single discourse.

This thinking is also taking hold on the international stage, as indicated by recent comments at the United Nations Conference on Trade and Industry:

> Integrating the acquisition of entrepreneurial competencies and "soft skills" such as creativity, initiative, and persuasion in the curriculum across all ages and subjects... [implies] a shift from a traditional emphasis in many education systems on evaluating the ideas of others to generating ideas oneself.[108]

We consider innovation to be a subset of creativity, and opportunity recognition to be a subset of innovation. If key drivers of innovation lie with those taking forward the entrepreneurship agenda,[109] and enterprise and entrepreneurship are seen to be about applied creativity,[110] this implies a need for the type of discussion presented at AUC. At a time when Sony acknowledges that it has a shortage of ideas,[111] and the *Wall Street Journal*'s most influential business thinker discusses relentless change and innovation, asking, "What happens now?,"[112] this type of discourse opens new channels of investigation.

Using the concept of diagonal thinking[113] and incorporating what is often termed left- and right-brain thinking into a single pedagogical stance enables recent trends and contextual decision making to be accommodated. We propose that it is important for the educator to frame his or her approach in a manner that provides learners with opportunities to become curious and learn to make new discoveries for themselves effectively.[114]

One barrier to divergent thinking is what we describe as "premature articulation," bringing a solution to bear before it has been fully researched in the broadest possible way (for example, by looking at the problems and their root causes, and not merely taking them at face value). We know that extended research takes time to assimilate in the brain, especially as extremely divergent perspectives may be represented. We also know that much of this thinking takes place subconsciously, and comes to light in a moment of realization, not so much rationally developed but more of a sudden insight. This infers time-related management and consideration that we see in figures 3.3 and 3.4.

These figures also illuminate the fact that many complex decision-making processes are initiated intuitively through the neurons in the limbic system. Our brains often only work out why they reached an answer after the event.[115] "Rather than actually guiding or controlling behavior, consciousness seems mainly to make sense of behavior after it is executed."[116] We have also considered what Gordon and Berger describe as "intelligent memory,"[117] seeing beyond the obvious and challenging one's own views of the world.

In developing the AUC presentation, we were acutely aware that we would be challenging accepted norms, and that the aspects of creativity and innovation we would be introducing needed to be effectively communicated. Further, as the fleeting nature of new understandings makes them difficult to capture, and even more difficult for the inexperienced learner to define, we had to ensure that there was sufficient understanding gained through experience, engagement, and emotional connection. We had to attempt to practice what we preach. For example, it may not have been immediately apparent to the delegates that the theoretical aspects were explained once personal and emotional understandings were established, a method that contradicts the classic model of explaining and telling at the outset, usually through the medium of the lecture. Many educators believe learners should simply be told the correct answer through "transmission teaching,"[118] so we set a scene where that was impossible, or at least difficult from the outset, by obtaining delegates' views within moments of opening the presentation. "The teacher is, after all, a role model for the students,"[119] and hence we had to make good effort to achieve this, despite the formal organizational setting.

Conclusion

Our chapter argues that much of the literature on enterprise and entrepreneurship is limited, as it has evolved from silos of expertise, especially from professors of business administration. Creativity and innovation require skills that enhance opportunity recognition by seeing outside of one's own silo and considering other specialists and knowledge. Capturing these and various other understandings may well provide more opportunities to intersect knowledge and educational processes than we have previously been aware of, but have not yet assembled in new ways or viewed from a holistic perspective.

We assert that pedagogies that employ design-based understanding, some of which were demonstrated at AUC, are a good fit with the needs of enterprise and entrepreneurship education, providing frameworks for "constructively aligned" assessment[120] and interdisciplinarity. The cognitive advantages include avoiding "premature articulation" when an idea has not been wholly formed,[121] and encouraging "glorious failures,"[122] where the use of appropriate cognitive processes within situational contexts informs the assessment, not merely final outcomes. Design-based pedagogy is a lesson of prototypes, not a lesson of final conclusions; it embraces human emotions and seeks out multiple solutions so that a range of alternatives is envisaged. It is teaching to prepare learners for an ever-changing environment, not a static one.

Turning ideas into action and progressively evaluating them in an iterative process of prototyping is an emulation of both the design thinker and the enterprising individual. We conclude by revisiting our original premise and ask: if creativity, innovation, and opportunity recognition were removed from our educational offerings, what would be left? Moreover, does enterprise start with business knowledge or creative capacity enhancement? Most important, if they ignore these questions, are entrepreneurial educators putting the cart before the horse?

Notes

1   Benson Honig, "Entrepreneurship Education: Towards a Model of Contingency-based Business Planning," *Academy of Management Learning and Education* 3 (2004): 258–73; Andy Penaluna and Kathryn Penaluna, "The Evidence So Far: Calling for Creative Industries Engagement with Entrepreneurship Education Policy and Development," in *Entrepreneurship and the Creative Economy: Process, Practice and Policy*, ed. Colette Henry and Anne De Bruin (Cheltenham: Edward Elgar, 2011), 50–78.

2   Amar Bhide, "Bootstrap Finance: The Art of Start Ups," *Harvard Business Review* 70, no. 6 (1992): 109–17.

3   Eric Ries, *The Lean Startup: How Today's Entrepreneurs Use Continuous Innovation to Create Radically Successful Businesses* (New York: Crown Business, 2011).

4   Welsh Assembly Government/Welsh Enterprise Educators Network, *Feasible, Desirable or What? An Investigation into the Feasibility of an Entrepreneurship Educators' Programme in Wales*, Cardiff: WEEN, 2010.

5   Welsh Assembly Government/Welsh Enterprise Educators Network, *Feasible, Desirable or What?*, 19.

6   QAA, *Quality Assurance Agency for Higher Education*, 2012a, http://www.qaa.ac.uk

7   QAA, *Enterprise and Entrepreneurship: A New Approach to Learning*, 2012, http://www.qaa.ac.uk/Newsroom/PressReleases/Pages/Enterprise-and-entrepreneurship-a-new-approach-to-learning.aspx

8   Christine Volkmann et al., *Educating the Next Wave of Entrepreneurs: Unlocking Entrepreneurial Capabilities to Meet the Global Challenges of the 21st Century—A Report of the Global Education Initiative*, Report for the Entrepreneurial Learning Initiative, 2009, http://www.gvpartners.com/web/pdf/WEF_EE_Full_Report.pdf

9   Jules L. Coleman, "Negative and Positive Positivism," *Journal of Legal Studies* 11 (1982): 139–64.

10 Egon G. Guaba and Yvonna S. Lincoln, "Competing Paradigms in Qualitative Research," in *Handbook of Qualitative Research*, ed. Norman K. Denzin and Yvonna S. Lincoln (London: Sage, 1994), 111; see also Andy Penaluna et al., *Entrepreneurial Education Needs Entrepreneurial Educators: Assessing Our Performance*, Proceedings of the 31st International Conference of the Institute for Small Business and Entrepreneurship, Belfast, November 2008, http://www.isbe.org.uk/APenaluna08

11 W. Ed McMullan and L.M. Gillin, "Entrepreneurship Education in the Nineties, Revisited," in *Entrepreneurship Education: A Global View*, eds. Robert H. Brockhaus et al., (Aldershot: Ashgate Publishing Ltd., 2001), 57–77.

12 Harry Matlay, "Entrepreneurship Education in UK Business Schools: Conceptual, Contextual and Policy Considerations," *Journal of Small Business and Enterprise Development* 12 (2005): 627–43.

13 CMI/NESTA, *Innovation for the Recovery: Enhancing Innovative Working Practices* (London: Chartered Management Institute/National Endowment for Science, Technology and the Arts, 2009); George Cox, *The Cox Review of Creativity in Business: Building on the UK's Strengths* (London: H.M. Treasury, 2005); Hartmut Esslinger, *A Fine Line: How Design Strategies are Shaping the Future of Business* (San Francisco: Jossey Bass, 2009); Veronica Hope-Halley, "A Fundamental Review of Business Education Is Overdue," *Times Online*, October 13, 2009, http://www.thetimes.co.uk/tto/career/article1839511.ece; Thomas Lockwood, ed., *Design Thinking: Integrating Innovation, Customer Experience and Brand Value* (New York: Allworth Press, 2010); Roger Martin, *The Design of Business: Why Design Thinking Is the Next Competitive Advantage* (Boston: Harvard Business Press, 2009); UNCTAD-MYM, *United Nations Conference on Trade and Development: Multi Year Expert Meeting on Enterprise Development Policies and Capacity-building in Science, Technology and Innovation*, Geneva, January 19–21, 2011, http://www.unctad.org/Templates/Page.asp?intItemID=5799&lang=1&print=1

14 Alex Byrne, "Interpretivism," *European Review of Philosophy* 3 (1998), http://web.mit.edu/abyrne/www/Interpretivism.html

15 Rupert Hall, "Senior UN Official to Work with University of Wales Trinity Saint David Swansea," *Wales Online*, October 28, 2013, http://www.walesonline.co.uk/business/business-news/senior-un-official-work-university-6245743.

16 European Commission, *Entrepreneurship Education: A Guide for Educators* (Brussels: Entrepreneurship and Social Economy Unit Directorate-General for Enterprise and Industry, 2013).

17 David Rae, Lynn Martin, Valerie Antcliff, and Paul Hannon, *The 2010 Survey of Enterprise and Entrepreneurship in Higher Education*, Proceedings of the 33rd ISBE Conference, London, November 2010, http://www.ncge.org.uk/publication/ISBE_Report.pdf, 3

18 "Oslo Agenda for Entrepreneurship Education in Europe." Agenda presented at the Entrepreneurship Education in Europe: Fostering Entrepreneurial Mindsets through Education and Learning Conference, Oslo, October 26–27, 2006, http://ec.europa.eu/enterprise/policies/sme/files/support_measures/training_education/doc/oslo_agenda_final_en.pdf, 3.

19 World Economic Forum, *Insight Report: Enhancing Europe's Competitiveness: Fostering Innovation-driven Entrepreneurship in Europe* (World Economic Forum, 2014), 5–8.

20 Jim Bell, Iian Callaghan, Dave Demick, and Fred Scharf, "Internationalising Entrepreneurship Education," *Journal of International Entrepreneurship* 2 (2004): 109–24.

21 Donald F. Kuratko, "The Emergence of Entrepreneurship Education: Development, Trends and Challenges," *Entrepreneurship Theory and Practice* 29 (2005): 583.

22 Jerome A. Katz, "The Chronology and Intellectual Trajectory of American Entrepreneurship Education," *Journal of Business Venturing* 18 (2003): 283–300; Kuratko, "The Emergence of Entrepreneurship Education."

23 Luke Pittaway and Corina Edwards, *Assessment: Examining Practice in Enterprise Education* (in review), 2012, http://eagleentrepreneur.files.wordpress.com/2012/01/lp-assessment-final-pre-review-edit.pdf

24 IEEC, *Proceedings of the International Entrepreneurship Educators Programme*, Cambridge University: Enterprise Educators UK and the National Council for Graduate Entrepreneurship, 2007, http://www.ieec.co.uk/2007/docs/Conference_Report.pdf

25 Mark A. Runco, *Divergent Thinking* (Norwood: Ablex Publishing Corporation, 1991); Andy Penaluna, Jackie Coates, and Kathryn Penaluna, "Creativity-Based Assessment and Neural Understandings: A Discussion and Case Study Analysis," *Education + Training* 52 (2010): 660–78.

26 Mark A. Runco, *Creativity Theories and Themes: Research, Development, and Practice* (San Diego: Elsevier Academic Books, 2007), 5.

27 David Kirby, "Entrepreneurship Education: Can Business Schools Meet the Challenge?" *Education + Training* 46 (2004): 510–19.

28 Robert Chia, "Teaching Paradigm Shifting in Management Education: University Business Schools and the Entrepreneurial Imagination," *Journal of Management Studies* 33 (1996): 409–28; Cecile Nieuwenhuizen and Darelle Groenewald, "Entrepreneurship Training and Education Needs as Determined by the Brain Preference Profiles of Successful Established Entrepreneurs" (paper presented at the Internationalizing Entrepreneurship Education and Training Conference, Naples, Italy, July 5–7, 2004); Alan Gibb, "Creating Conducive Environments for Learning and Entrepreneurship: Living with, and Dealing with, Creating and Enjoying Uncertainty and Complexity," *Entrepreneurship Theory and Practice* 16 (2002): 135–48; Alan Gibb, *Towards the Entrepreneurial University: Entrepreneurship Education as a Lever for Change*, Policy Paper 003 (Birmingham: National Council for Graduate Entrepreneurship, 2005); Andy Penaluna and Kathryn Penaluna, "Business Paradigms in Einstellung: Harnessing Creative Mindsets, a Creative Industries Perspective," *Journal of Small Business and Entrepreneurship* 21 (2008): 231–50.

29 Jarna Heinonen et al., "The Role of Creativity in Opportunity Search and Business Idea Creation," *Education + Training* 53 (2011): 650–72; Jonathan Scott et al., "'What's the Impact?' The Potential for Experiential Learning to Achieve More Effective Learning Outcomes in Entrepreneurship Education Modules" (paper presented at Institute for Small Business and Entrepreneurship Conference, Dublin, 2012); Jane Silver and Claire MacLean, "Fostering an Entrepreneurial Mindset: A Pan-university Holistic Approach" (paper presented at Internationalizing Entrepreneurship Education and Training Conference, Gdansk, Poland, 2007).

30 *The Impact of Culture on Creativity* (study prepared for the European Commission's Directorate-General for Education and Culture, 2009), 115.

31 Roger Martin, *The Design of Business: Why Design Thinking is the Next Competitive Advantage* (Boston: Harvard Business School Press, 2009).

32 Stuart Cunningham, "Creative Enterprises," in *Creative Industries*, ed. John Hartley (Maiden: Blackwell Publishing, 2005), 7, 282–98.

33 Tim Brown, *Change by Design: How Design Thinking Transforms Organizations and Inspires Innovation* (New York: Harper Business, 2009).

34 Deniz Ucsbasaran, Paul Westhead, and Mike Wright, "The Focus of Entrepreneurial Research: Contextual and Process Issues," *Entrepreneurship Theory and Practice* 13 (2001): 58.

35 Colette Henry, Frances Hill, and Claire Leitch, "Entrepreneurship Education and Training: Can Entrepreneurship Be Taught? Part 1," *Education and Training* 47 (2005): 99.

36 Eugene Fregetto, "Do Entrepreneurial Inclined Students Learn More from Simulations?" Entrepreneurship Education Track, United States Association for Small Business and Entrepreneurship National Conference, Tucson, Arizona, 2006.

37 Simon 1981, quoted in Kung Wong Lau, "Creativity Training in Higher Design Education," *Design Journal* 12 (2009): 155.

38 Gerald Zaltman, *How Customers Think: Essential Insights into the Mind of the Market*, (Boston: Harvard Business School Press, 2003), xii.

39 Howard Gardner, *Art, Mind and Brain: A Cognitive Approach to Creativity* (New York: Basic Books, 1982); Jose Gomez, "What Do We Know About Creativity?" *Journal of Effective Teaching* 7 (2007): 31–43; E. Paul Torrance, "Can We Teach Children to Think Creatively?" *Journal of Creative Behavior* 6 (1972): 114–43; E. Paul Torrance, "Creative Teaching Makes a Difference," in *Creativity: Its Educational Implications*, ed. John Curtis Gowan, Joe Khatena, and E. Paul Torrance (Dubuque: Kendall/Hunt, 1981), 99–108.

40 Joseph Schumpeter, *The Theory of Economic Development* (Cambridge: Harvard University Press, 1934).

41 Sarah-Jayne Blakemore and Uta Frith, *The Learning Brain: Lessons for Education* (Malden, Oxford, and Victoria: Blackwell Publishing, 2005).

42 American University in Cairo, "Entrepreneurship and Innovation: Shaping the Future of Egypt" conference (http://www.aucegypt.edu/Business/newsroom/Pages/AUC-Research-Conference.aspx, April 17–19, 2012; for the full program and details of the keynote presentation, see http://conf.aucegypt.edu/Conferences/ConfHome.aspx?Conf=EIResearchconf&Title=Abstracts+and+Presentations)

43 Robert Cooper, *Winning at New Products: Accelerating the Process from Idea to Launch*, 3rd ed. (New York: Basic Books, 2001); John Bessant and Joe Tidd, *Innovation and Entrepreneurship* (Chichester: Wiley, 2007).

44 Daniel Pink, The Surprising Truth about What Motivates Us, (Edinburgh: Cannongate Books, 2011)

45 Robert J. Sternberg and Todd I. Lubart, "An Investment Theory of Creativity and Its Development," *Human Development* 34 (1991): 1–31.

46 James R. Flynn, *What Is Intelligence? Beyond the Flynn Effect* (New York: Cambridge University Press, 2007).

47 Kyung Hee Kim, "The Creativity Crisis: The Decrease in Creative Thinking Score on the Torrance Tests of Creative Thinking," *Creativity Research Journal* 23 (2011): 289–90.

48 David Pink, *A Whole New Mind: Why Right-Brainers Will Rule the Future*, new ed. (London: Marshall Cavendish, 2008).

49 Pink, *A Whole New Mind*.

50 Donald A. Schön, *The Reflective Practioner: How Professionals Think in Action* (New York: Basic Books, 1983); David Kolb, *Experiential Learning: Experience as the Source of Learning Development* (Englewood Cliffs, NJ: Prentice Hall, 1984).

51 Pink, *A Whole New Mind*.

52 Paul Thagard and Cameron Shelley, *Abductive Reasoning: Logic, Visual Thinking, and Coherence* (Waterloo: University of Waterloo), http://cogsci.uwaterloo.ca/Articles/Pages/%7FAbductive.html

53 Yun Peng and James Reggia, *Abductive Inference Models for Diagnostic Problem Solving* (New York: Springer, 1990).

54 Henk G. Schmidt, "Problem-based Learning: Rationale and Description," *Medical Education* 17 (1983): 11–16; John Savery, "Overview of Problem-based Learning: Definitions and Distinctions," *Interdisciplinary Journal of Problem-based Learning* 1 (2006), Article 3; Penaluna et al., "Creativity-Based Assessment."

55 Samy A. Azer, "Problem-based Learning: A Critical Review of Its Educational Objectives and the Rationale for Its Use," *Saudi Medical Journal* 22 (2001): 299–305.

56 Ap Dijksterhuis and Teun Meurs, "Where Creativity Resides: The Generative Power of Unconscious Thought," *Consciousness and Cognition* 15 (2006): 135–46.

57 Mitchell Ash, *Gestalt Psychology in German Culture, 1890–1967* (Cambridge: Cambridge University Press, 1998); Michael Wertheimer, "Experimentelle Studien über das Sehen von Bewegung," *Zeitschrift für Psychologie* 61 (1912): 161–265.

58 Michotte 1991, quoted in Georges Thines, Alan Costall, and Geroge Butterworth, eds., *Michotte's Experimental Phenomenology of Perception* (Hillsdale: Erlbaum, 1991), 220; see also Spokane Falls Graphic Design, *The Gestalt Principles*, http://graphicdesign.spokanefalls.edu/tutorials/process/gestaltprinciples/gestaltprinc.htm

59 Alan Gibb, "Entrepreneurship and Small Business Management: Can We Afford to Neglect Them in the Twenty-first-century Business School?" *British Journal of Management* 7 (1996): 309–21; Susanne Ollila and Karen Williams-Middleton, "The Venture Creation Approach: Integrating Entrepreneurial Education and Incubation at the University," *International Journal of Entrepreneurship and Innovation Management* 13 (2011): 161–78.

60 Andy Penaluna and Kathryn Penaluna, "Business Paradigms in Einstellung"; Luke Pittaway and Jason Cope, "Entrepreneurship Education: A Systematic Review of the Evidence," *International Small Business Journal* 25 (2007): 477–506; Luke Pittaway and Jason Cope, "Simulating Entrepreneurial Learning: Integrating Experiential and Collaborative Approaches to Learning," *Management Learning* 3 (2007): 211–33; Scott et al., "'What's the Impact?'"

61 Ollila and Williams-Middleton, "The Venture Creation Approach."

62 Silver and MacLean, "Fostering an Entrepreneurial Mindset"; Scott et al., "'What's the Impact?'"

63 Andy Penaluna et al., "Creativity-based Assessment and Neural Understandings"; Pittaway and Edwards, *Assessment*.

64 Scott et al., "'What's the Impact?'"

65 John Biggs, *Teaching for Quality Learning at University* (Berkshire: Open University Press, 2003).

66 Torrance, "Can We Teach Children to Think Creatively?"

67 Teresa M. Amabile et al., "Creativity under the Gun," *Harvard Business Review* 80 (2002): 52–61.

68 APA Presidential Task Force on Psychology in Education, "Learner-centered Psychological Principles: Guidelines for School Redesign and Reform" (American Psychological Association and the Mid-Continent Regional Education Lab: Washington, DC, 1993).

69 Robert J. Sternberg and Wendy M. Williams, *How to Develop Student Creativity* (Alexandria: Association of Supervision and Curriculum Development, 1996).

70 Torrance, "Creative Teaching Makes a Difference."

71 Merle Karnes et al., *Factors Associated with Underachievement and Overachievement of Intellectually Gifted Children* (Champaign, IL: Champaign Community Unit Schools, 1961).

72 Jami J. Shah, Noe Vargas-Hernandez, and Steve Smith, "Metrics for Measuring Ideation Effectiveness," *Design Studies* 24 (2003:) 111–34.

73 Franklin Institute, *The Franklin Institute Awards: History of and Facts about the Awards*, http://www.fi.edu/franklinawards/about.html, 2012.

74 Penaluna et al., "Creativity-based Assessment and Neural Understandings."

75 Kounios, quoted in Jonah Lehrer, "The Eureka Hunt: Why Do Good Ideas Come to Us When They Do?" *The New Yorker*, July 28, 2008, 42.

76 John Kounios et al., "The Prepared Mind: Neural Activity Prior to Problem Presentation Predicts Subsequent Solution by Sudden Insight," *Psychological Science* 17 (2006): 882–90.

77 Kounios et al., "The Prepared Mind," 882.

78 Zaltman, *How Customers Think*, 8.

79 Theodore W. Berger et al., "A Cortical Neural Prosthesis for Restoring and Enhancing Memory," *Journal of Neural Engineering* 8 (2011): 1–11.

80 Alicia C. Martinez et al., *Global Entrepreneurship Monitor Special Report: A Global Perspective on Entrepreneurship Education and Training* (Babson Park, MA: Babson College, 2010).

81 NESTA, *Developing Entrepreneurial Graduates: Putting Entrepreneurship at the Centre of Higher Education* (London: NESTA, 2008), http://ncee.org.uk/wpcontent/uploads/2014/06/developing_entrepreneurial_graduates.1.pdf

82 William G. Huitt and John Hummel, "An Introduction to Operant (instrumental) Conditioning," *Educational Psychology Interactive* (Valdosta, GA: Valdosta State University, 1997), http://www.edpsycinteractive.org/topics/behavior/operant.html.

83 Ivan Pavlov, *Conditioned Reflexes* (London: Oxford University Press, 1927).

84 For a more detailed discussion see Blakemore and Frith, *The Learning Brain*.

85 Raymond P. Kesner, *Behavioral Functions of the CA3 Subregion of the Hippocampus: Learning and Memory* (Cold Spring Harbor, NY: Cold Spring Harbor Laboratory Press, 2007), 1, http://www.learnmem.org/cgi/doi/10.1101/lm.688207.

86 Blakemore and Frith, *The Learning Brain*, 133.

87 Joseph S. Renzulli, "A General Theory for the Development of Creative Productivity through the Pursuit of Ideal Acts of Learning," *Gifted Child Quarterly* 36 (1992): 170–82.

88 Mehmet A. Gülpinar, "The Principles of Brain-based Learning and Constructivist Models in Education," *Educational Sciences: Theory & Practice* 5 (2005): 1.

89 For a detailed discussion see Kounios et al., "The Prepared Mind."

90 Blakemore and Frith, *The Learning Brain*, 179.

91 QAA, *Enterprise and Entrepreneurship*.

92 Daniel Fasko, "Education and Creativity," *Creativity Research Journal* 13 (2000–2001): 317–27.

93 Sidney Parnes and Ruth Noller, "Applied Creativity: The Creative Studies Project," *Journal of Creative Behavior* 6 (1972): 164–86.

94 Amabile et al., "Creativity under the Gun," 52.

95 IPA, *Institute of Practitioners in Advertising: Diagonal Thinking* (London: IPA, 2007), http://www.ipa.co.uk/Page/diagonal-thinking-introduction.

96 See Diagonal Thinking, "Diagonal Thinking Self-Assessment Test," http://www.diagonalthinking.co.uk.

97 Lehrer, "The Eureka Hunt," 42.

98 For a fuller discussion and a case study of this process in operation, see Penaluna et al., "Creativity-Based Assessment and Neural Understandings."

99 Tim Wilson, *A Review of Business–University Collaboration* (London, 2012), 53.

100 QAA, *Art Design Subject Benchmark Statement*, 6.

101 Biggs, *Teaching for Quality Learning*.

102 For a fuller account, see Penaluna and Penaluna, "The Evidence So Far."

103 Lev Semenovich Vygotsky, *Mind in Society: The Development of Higher Psychological Processes* (Harvard University Press, 1978).

104 For a fuller discussion on this diagram see: All Party Parliamentary Group for Micro Businesses, *An Education System Fit for an Entrepreneur* (London: House of Commons/UK Parliament, 2014).

105 Teresa M. Amabile, *The Social Psychology of Creativity* (New York: Springer-Verlag, 1983).

106 Geert Hofstede, *Culture's Consequences: International Differences in Work-related Values*, 2nd ed. (Beverly Hills: Sage, 1984).

107 Biggs, *Teaching for Quality Learning*.

108 UNCTAD-MYM, *United Nations Conference on Trade and Development*, 7.

109 European Commission, *Education for Entrepreneurship: Making Progress in Promoting Entrepreneurial Attitudes and Skills through Primary and Secondary Education* (Brussels: European Commission, 2004).

110 OECD, *Evaluation of Programmes Concerning Education for Entrepreneurship*, report by the OECD Working Party on SMEs and Entrepreneurship, 2009; David Rae, *Entrepreneurship: From Opportunity to Action* (Basingstoke: Palgrave McMillan, 2007).

111 Bryan Gruley and Cliff Edwards, "Sony: All Chewed Up," *Management Today* (February 2012): 51–56.

112 Gary Hamel, *What Matters Now: How to Win in a World of Relentless Change, Ferocious Competition and Unstoppable Innovation* (San Francisco: Jossey Bass, 2012).

113 See Diagonal Thinking, "Diagonal Thinking Self-assessment Test."

114 Penaluna et al., "Creativity-based Assessment and Neural Understandings."

115 Erik Du Plessis, *The Advertised Mind: Ground-breaking Insights into How Our Brains Respond to Advertising* (London: Kogan Page, 2005); Erik Du Plessis, *The Branded Mind: What Neuroscience Really Tells Us about the Puzzle of the Brain and the Brand* (London: Kogan Page, 2011).

116 Lowenstein, quoted in Zaltman, *How Customers Think*, 10.

117 Barry Gordon and Lisa Berger, *Intelligent Memory: Exercise Your Mind and Make Yourself Smarter* (London: Vermillion, 2003).
118 Clive Beck and Clare Kosnik, *Innovations in Teacher Education: A Social Constructivist Approach* (Albany: State University of New York Press, 2006).
119 Runco, *Creativity Theories and Themes*, 189.
120 Biggs, *Teaching for Quality Learning*.
121 Penaluna et al., "Creativity-based Assessment and Neural Understandings."
122 Andy Penaluna and Kathryn Penaluna, *Entrepreneurship for Artists and Designers in Higher Education*, proceedings of the Internationalizing Entrepreneurship Education and Training Conference, Guildford, July 10–13, 2005, http://www. intent-conference.com/structure_default/ePilot40.asp?G=621&A=1.

Bibliography

All Party Parliamentary Group for Micro Businesses. *An Education System Fit for an Entrepreneur*. London: House of Commons/UK Parliament, 2014.
Amabile, Teresa M. *The Social Psychology of Creativity*. New York: Springer-Verlag, 1983.
———. Constance N. Hadley, and Steven J. Kramer. "Creativity under the Gun." *Harvard Business Review* 80 (2002): 52–61.
American University in Cairo. "Entrepreneurship and Innovation: Shaping the Future of Egypt." Conference, April 17–19, 2012. http://www.aucegypt.edu/ Business/newsroom/Pages/AUC-Research-Conference.aspx
APA Presidential Task Force on Psychology in Education. "Learner-centered Psychological Principles: Guidelines for School Redesign and Reform." Washington, DC: American Psychological Association and the Mid-Continent Regional Education Lab, 1993.
Ash, Mitchell. *Gestalt Psychology in German Culture, 1890–1967*. Cambridge: Cambridge University Press, 1998.
Azer, Samy A. "Problem-based Learning: A Critical Review of Its Educational Objectives and the Rationale for Its Use." *Saudi Medical Journal* 22 (2001): 299–305.
Beck, Clive, and Clare Kosnik. *Innovations in Teacher Education: A Social Constructivist Approach*. Albany: State University of New York Press, 2006.
Bell, Jim, Iian Callaghan, Dave Demick, and Fred Scharf. "Internationalising Entrepreneurship Education." *Journal of International Entrepreneurship* 2 (2004): 109–24.
Berger, Theodore W., Robert E. Hampson, Dong Song, Anushka Goonawardena, Vasilis Z. Mamarelis, and Sam A. Deadwyler. "A Cortical Neural Prosthesis for Restoring and Enhancing Memory." *Journal of Neural Engineering* 8 (2011): 1–11.
Bessant, John, and Joe Tidd. *Innovation and Entrepreneurship*. Chichester: Wiley, 2007.
Bhide, Amar. "Bootstrap Finance: The Art of Start Ups." *Harvard Business Review* 70, no. 6 (1992): 109–17.
Biggs, John. *Teaching for Quality Learning at University*. Berkshire: Open University Press, 2003.
Blakemore, Sarah-Jayne, and Uta Frith. *The Learning Brain: Lessons for Education*. Malden, Oxford, and Victoria: Blackwell Publishing, 2005.
Brown, Tim. *Change by Design: How Design Thinking Transforms Organizations and Inspires Innovation*. New York: Harper Business, 2009.
Byrne, Alex. "Interpretivism." *European Review of Philosophy* 3 (1998). http://web.mit. edu/abyrne/www/Interpretivism.html

Chia, Robert. "Teaching Paradigm Shifting in Management Education: University Business Schools and the Entrepreneurial Imagination." *Journal of Management Studies* 33 (1996): 409–28.

CMI/NESTA. *Innovation for the Recovery: Enhancing Innovative Working Practices.* London: Chartered Management Institute/National Endowment for Science, Technology and the Arts, 2009.

Coleman, Jules L. "Negative and Positive Positivism." *Journal of Legal Studies* 11 (1982): 139–64.

Cooper, Robert. *Winning at New Products: Accelerating the Process from Idea to Launch.* 3rd ed. New York: Basic Books, 2001.

Cox, George. *The Cox Review of Creativity in Business: Building on the UK's Strengths.* Report prepared for the Chancellor of the Exchequer. London: HM Treasury, 2005.

Cunningham, Stuart. "Creative Enterprises." In *Creative Industries,* edited by John Hartley, 282–98. Maiden: Blackwell Publishing, 2005.

Dijksterhuis, Ap, and Teun Meurs. "Where Creativity Resides: The Generative Power of Unconscious Thought." *Consciousness and Cognition* 15 (2006): 135–46.

Du Plessis, Erik. *The Advertised Mind: Ground-breaking Insights into How Our Brains Respond to Advertising.* London: Kogan Page, 2005.

———. *The Branded Mind: What Neuroscience Really Tells Us about the Puzzle of the Brain and the Brand.* London: Kogan Page, 2011.

Esslinger, Hartmut. *A Fine Line: How Design Strategies are Shaping the Future of Business.* San Francisco: Jossey Bass, 2009.

European Commission. *Education for Entrepreneurship: Making Progress in Promoting Entrepreneurial Attitudes and Skills through Primary and Secondary Education.* Brussels: European Commission, 2004.

———. *Entrepreneurship Education: A Guide for Educators.* Brussels: Entrepreneurship and Social Economy Unit Directorate-General for Enterprise and Industry, 2013.

Fasko, Daniel. "Education and Creativity." *Creativity Research Journal* 13 (2000–2001): 317–27.

Flynn, James. R. *What Is Intelligence? Beyond the Flynn Effect.* New York: Cambridge University Press, 2007.

Franklin Institute. *The Franklin Institute Awards: History of and Facts about the Awards.* http://www.fi.edu/franklinawards/about.html, 2012.

Fregetto, Eugene. "Do Entrepreneurially-inclined Students Learn More from Simulations?" Entrepreneurship Education Track, United States Association for Small Business and Entrepreneurship National Conference, Tucson, Arizona, 2006.

Gardner, Howard. *Art, Mind and Brain: A Cognitive Approach to Creativity.* New York: Basic Books, 1982.

Gibb, Alan. "Creating Conducive Environments for Learning and Entrepreneurship: Living with, and Dealing with, Creating and Enjoying Uncertainty and Complexity." *Entrepreneurship Theory and Practice* 16 (2002): 135–48.

———. "Entrepreneurship and Small Business Management: Can We Afford to Neglect Them in the Twenty-first Century Business School?" *British Journal of Management* 7 (1996): 309–21.

———. *Towards the Entrepreneurial University: Entrepreneurship Education as a Lever for Change.* Policy Paper 003. Birmingham: National Council for Graduate Entrepreneurship, 2005.

Gomez, Jose. "What Do We Know About Creativity?" *Journal of Effective Teaching* 7 (2007): 31–43.

Gordon, Barry, and Lisa Berger. *Intelligent Memory: Exercise Your Mind and Make Yourself Smarter.* London: Vermillion, 2003.

Gruley, Bryan, and Cliff Edwards. "Sony: All Chewed Up." *Management Today* (February 2012): 51–56.

Guba, Egon G., and Yvonna S. Lincoln. "Competing Paradigms in Qualitative Research." In *Handbook of Qualitative Research*, edited by Norman K. Denzin and Yvonna S. Lincoln, 104–17. London: Sage, 1994.

Gülpinar, Mehmet A. "The Principles of Brain-based Learning and Constructivist Models in Education." *Educational Sciences: Theory & Practice* 5 (2005): 299–306.

Hall, Rupert. "Senior UN Official to Work with University of Wales Trinity Saint David Swansea." *Wales Online*, October 28, 2013. http://www.walesonline.co.uk/business/business-news/senior-un-official-work-university-6245743

Hamel, Gary. *What Matters Now: How to Win in a World of Relentless Change, Ferocious Competition and Unstoppable Innovation.* San Francisco: Jossey Bass, 2012.

Heinoven, Jaran, Ulla Hytti, and Pekka Stenholm. "The Role of Creativity in Opportunity Search and Business Idea Creation." *Education + Training* 53 (2011): 659–72.

Henry, Colette, Frances Hill, and Claire Leitch. "Entrepreneurship Education and Training: Can Entrepreneurship Be Taught? Part 1." *Education and Training* 47 (2005): 98–111.

Hofstede, Geert. *Culture's Consequences: International Differences in Work-Related Values.* 2nd ed. Beverly Hills: Sage, 1984.

Honig, Benson. "Entrepreneurship Education: Towards a Model of Contingency-based Business Planning." *Academy of Management Learning and Education* 3 (2004): 258–73.

Hope-Halley, Veronica. "A Fundamental Review of Business Education is Overdue." *Times Online*, October 13, 2009. http://www.thetimes.co.uk/tto/career/article1839511.ece

Huitt, William G., and John Hummel. "An Introduction to Operant (Instrumental) Conditioning." *Educational Psychology Interactive*. Valdosta, GA: Valdosta State University, 1997. http://www.edpsycinteractive.org/topics/behavior/operant.html.

IEEC. *Proceedings of the International Entrepreneurship Educators Programme.* Cambridge University: Enterprise Educators UK and the National Council for Graduate Entrepreneurship, 2007. http://www.ieec.co.uk/2007/docs/Conference_Report.pdf

The Impact of Culture on Creativity. Study prepared for the European Commission's Directorate-General for Education and Culture, 2009.

IPA. *Institute of Practitioners in Advertising: Diagonal Thinking.* London: IPA, 2007. http://www.ipa.co.uk/Page/diagonal-thinking-introduction.

Karnes, Merle, G.F. McCoy, R.R Zehrbach, J.P. Wollersheim, H.F. Clarizo, L. Costin, and L.S. Stanley. *Factors Associated with Underachievement and Overachievement of Intellectually Gifted Children.* Champaign, IL: Champaign Community Unit Schools, 1961.

Katz, Jerome A. "The Chronology and Intellectual Trajectory of American Entrepreneurship Education." *Journal of Business Venturing* 18 (2003): 283–300.

Kesner, Raymond P. *Behavioral Functions of the CA3 Subregion of the Hippocampus: Learning and Memory.* Cold Spring Harbor, NY: Cold Spring Harbor Laboratory Press, 2007. http://www.learnmem.org/cgi/doi/10.1101/lm.688207

Kim, Kyung Hee. "The Creativity Crisis: The Decrease in Creative Thinking Score on the Torrance Tests of Creative Thinking." *Creativity Research Journal* 23 (2011): 285–95.

Kirby, David. "Entrepreneurship Education: Can Business Schools Meet the Challenge?" *Education + Training* 46 (2004): 510–19.

Kolb, David. *Experiential Learning: Experience as the Source of Learning Development.* Englewood Cliffs, NJ: Prentice Hall, 1984.

Kounios, John, Jennifer L. Frymiare, Edward M. Bowden, Jessica I. Fleck, Karuna Subramaniam, Todd B. Parish, and Mark Jung-Beeman. "The Prepared Mind: Neural Activity Prior to Problem Presentation Predicts Subsequent Solution by Sudden Insight." *Psychological Science* 17 (2006): 882–90.

Kuratko, Donald F. "The Emergence of Entrepreneurship Education: Development, Trends and Challenges." *Entrepreneurship Theory and Practice* 29 (2005): 577–98.

Lau, Kung Wong. "Creativity Training in Higher Design Education." *Design Journal* 12 (2009): 153–69.

Lehrer, Jonah. "The Eureka Hunt: Why Do Good Ideas Come to Us When They Do?" *The New Yorker*, July 28, 2008, 40–45.

Lockwood, Thomas, ed. *Design Thinking: Integrating Innovation, Customer Experience and Brand Value.* New York: Allworth Press, 2010.

Martin, Roger. *The Design of Business: Why Design Thinking Is the Next Competitive Advantage.* Boston: Harvard Business Press, 2009.

Martinez, Alicia C., Jonathan Levie, Donna J. Kelley, J. Rognvaldur, R.J. Saedmundsson, and Thomas Schott. *Global Entrepreneurship Monitor Special Report: A Global Perspective on Entrepreneurship Education and Training.* Babson Park, MA: Babson College, 2010.

Matlay, Harry. "Entrepreneurship Education in UK Business Schools: Conceptual, Contextual and Policy Considerations." *Journal of Small Business and Enterprise Development* 12 (2005): 627–43.

McMullan, W.E., and L.M. Gillin. "Entrepreneurship Education in the Nineties, Revisited." In *Entrepreneurship Education: A Global View*, edited by R.H. Brockhaus, G.E. Hill, H. Klandt, and H.P. Welch, 57–77. Aldershot: Ashgate Publishing Ltd., 2001.

NESTA. *Developing Entrepreneurial Graduates: Putting Entrepreneurship at the Centre of Higher Education.* London: NESTA, 2008. http://ncee.org.uk/wp-content/uploads/2014/06/developing_entrepreneurial_graduates.1.pdf

Nieuwenhuizen, Cecile, and Darelle Groenewald. "Entrepreneurship Training and Education Needs as Determined by the Brain Preference Profiles of Successful Established Entrepreneurs." Paper presented at the Internationalizing Entrepreneurship Education and Training Conference, Naples, Italy, July 5–7, 2004.

OECD. *Evaluation of Programmes Concerning Education for Entrepreneurship.* Report by the OECD Working Party on SMEs and Entrepreneurship, 2009.

———. "LEED Programme (Local Economic and Employment Development)." http://www.oecd.org/cfe/leed/entrepreneurship360-skills-entrepreneurship.htm.

Ollila, Susanne, and Karen Williams-Middleton. "The Venture Creation Approach: Integrating Entrepreneurial Education and Incubation at the University." *International Journal Entrepreneurship and Innovation Management* 13 (2011): 161–78.

"Oslo Agenda for Entrepreneurship Education in Europe." Agenda presented at the Entrepreneurship Education in Europe: Fostering Entrepreneurial Mindsets

through Education and Learning Conference, Oslo, October 26–27, 2006. http://
ec.europa.eu/enterprise/policies/sme/files/support_measures/training_education/
doc/oslo_agenda_final_en.pdf.

Parnes, Sidney, and Ruth Noller. "Applied Creativity: The Creative Studies Project."
Journal of Creative Behavior 6 (1972): 164–86.

Pavlov, Ivan. *Conditioned Reflexes.* London: Oxford University Press, 1927.

Penaluna, Andy, Simon Brown, David Gibson, Colin Jones, and Kathryn Penaluna.
*Entrepreneurial Education needs Entrepreneurial Educators: Assessing Our
Performance.* Proceedings of the 31st International Conference of the Institute for
Small Business and Entrepreneurship, Belfast, 2008.

Penaluna, Andy, Jackie Coates, and Kathryn Penaluna. "Creativity-Based Assessment
and Neural Understandings: A Discussion and Case Study Analysis." *Education +
Training* 52 (2010): 660–78.

Penaluna, Andy, and Kathryn Penaluna. "Business Paradigms in Einstellung:
Harnessing Creative Mindsets, a Creative Industries Perspective." *Journal of
Small Business and Entrepreneurship* 21 (2008): 231–50.

———. *Entrepreneurship for Artists and Designers in Higher Education.* Proceedings of
the Internationalizing Entrepreneurship Education and Training Conference,
Guildford, July 10–13, 2005. http://www.intent-conference.com/structure_
default/ePilot40.asp?G=621&A=1.

———. "The Evidence So Far: Calling for Creative Industries Engagement with
Entrepreneurship Education Policy and Development." In *Entrepreneurship and
the Creative Economy: Process, Practice and Policy,* edited by Colette Henry and Anne
De Bruin, 50–78. Cheltenham: Edward Elgar, 2011.

Penaluna, Kathryn, Andy Penaluna, and Colin Jones. "The Context of Enterprise Education: Insights into Current Practices." *Industry and Higher Education* 26 (2012): 163–75.

Peng, Yun, and James Reggia. *Abductive Inference Models for Diagnostic Problem Solving.*
New York: Springer, 1990.

Pink, David. *A Whole New Mind: Why Right-brainers Will Rule the Future.* New ed.
London: Marshall Cavendish, 2008.

Pink, David. *RSA Animate-Drive: The Surprising Truth about What Motivates Us.* Royal
Society of Arts, 2010. http://www.youtube.com/watch?v=u6XAPnuFjJc

Pittaway, Luke, and Jason Cope. "Entrepreneurship Education: A Systematic Review
of the Evidence." *International Small Business Journal* 25 (2007): 477–506.

———. "Simulating Entrepreneurial Learning: Integrating Experiential and
Collaborative Approaches to Learning." *Management Learning* 3 (2007): 211–33.

Pittaway, Luke, and Corina Edwards. *Assessment: Examining Practice in Enterprise
Education* (in review), 2012. http://eagleentrepreneur.files.wordpress.
com/2012/01/lp-assessment-final-pre-review-edit.pdf

QAA. *Art Design Subject Benchmark Statement.* Gloucester, 2008. http://www.qaa.
ac.uk/Publications/InformationAndGuidance/Pages/Subject-benchmark-
statement-Art-and-design-.aspx.

———. *Enterprise and Entrepreneurship: A New Approach to Learning.* 2012.
http://www.qaa.ac.uk/Newsroom/PressReleases/Pages/Enterprise-and-
entrepreneurship-a-new-approach-to-learning.aspx

———. *Quality Assurance Agency for Higher Education.* 2012. http://www.qaa.ac.uk.

Rae, David. *Entrepreneurship: From Opportunity to Action.* Basingstoke: Palgrave
McMillan, 2007.

Rae, David, Lynn Martin, Valerie Antcliff, and Paul Hannon. *The 2010 Survey of Enterprise and Entrepreneurship in Higher Education*. Proceedings of the 33rd ISBE Conference, London, November 2010. http://www.ncge.org.uk/publication/ISBE _Report.pdf

Reis, Eric. *The Lean Startup: How Today's Entrepreneurs Use Continuous Innovation to Create Radically Successful Businesses*. New York: Crown Business, 2011.

Renzulli, Joseph S. "A General Theory for the Development of Creative Productivity through the Pursuit of Ideal Acts of Learning." *Gifted Child Quarterly* 36 (1992): 170–82.

Runco, Mark A. *Creativity Theories and Themes: Research, Development, and Practice*. San Diego: Elsevier Academic Books, 2007.

——. *Divergent Thinking*. Norwood: Ablex Publishing Corporation, 1991.

Savery, John. "Overview of Problem-based Learning: Definitions and Distinctions." *Interdisciplinary Journal of Problem-based Learning* 1 (2006), Article 3.

Schmidt, Henk G. "Problem-based Learning: Rationale and Description." *Medical Education* 17 (1983): 11–16.

Schön, Donald A. *The Reflective Practioner: How Professionals Think in Action*. New York: Basic Books, 1983.

Scott, Jonathan, Andy Penaluna, John Thompson, and David Brooksbank. "'What's the Impact?' Potential for Experiential Learning to Achieve More Effective Learning Outcomes in Entrepreneurship Education Modules." Paper presented at the Institute for Small Business and Entrepreneurship Conference, Dublin, 2012.

Schumpeter, Joseph. *The Theory of Economic Development*. Cambridge: Harvard University Press, 1934.

Shah, Jami J., Noe Vargas-Hernandez, and Steve Smith. "Metrics for Measuring Ideation Effectiveness." *Design Studies* 24 (2003): 111–34.

Silver, Jane, and Claire MacLean. "Fostering an Entrepreneurial Mindset: A Pan-university Holistic Approach." Paper presented at Internationalizing Entrepreneurship Education and Training Conference, Gdansk, Poland, 2007.

Spokane Falls Graphic Design. *The Gestalt Principles*. http://graphicdesign.spokanefalls.edu/tutorials/process/gestaltprinciples/gestaltprinc.htm.

Sternberg, Robert J., and Todd I. Lubart. "An Investment Theory of Creativity and Its Development." *Human Development* 34 (1991): 1–31.

Sternberg, Robert J., and Wendy M. Williams. *How to Develop Student Creativity*. Alexandria: Association of Supervision and Curriculum Development, 1996.

Thagard, Paul, and Cameron Shelley. *Abductive Reasoning: Logic, Visual Thinking, and Coherence*. Waterloo: University of Waterloo. http://cogsci.uwaterloo.ca/Articles/Pages/%7FAbductive.html

Thines, Georges, Alan Costall, and Geroge Butterworth, eds. *Michotte's Experimental Phenomenology of Perception*. Hillsdale: Erlbaum, 1991.

Torrance, E. Paul. "Can We Teach Children to Think Creatively?" *Journal of Creative Behavior* 6 (1972): 114–43.

——. "Creative Teaching Makes a Difference." In *Creativity: Its Educational Implications*, edited by John Curtis Gowan, Joe Khatena and E. Paul Torrance. 2nd ed., 99–108. Dubuque: Kendall/Hunt, 1981.

Ucsbasaran, Deniz, Paul Westhead, and Mike Wright. "The Focus of Entrepreneurial Research: Contextual and Process Issues." *Entrepreneurship Theory and Practice* 13 (2001): 13–17.

UNCTAD-MYM. *United Nations Conference on Trade and Development: Multi Year Expert Meeting on Enterprise Development Policies and Capacity-building in Science, Technology and Innovation.* Geneva, January 19–21, 2011. http://www.unctad.org/Templates/Page.asp?intItemID=5799&lang=1&print=1

Volkmann, Christine, Karen Wilson, Steve Mariotti, Daniel Rabuzzi, and Shailendra Vyakarnam. *Educating the Next Wave of Entrepreneurs: Unlocking Entrepreneurial Capabilities to Meet the Global Challenges of the 21st Century—A Report of the Global Education Initiative.* Report for the Entrepreneurial Learning Initiative, 2009. http://www.gvpartners.com/web/pdf/WEF_EE_Full_Report.pdf

Vygotsky, Lev Semenovich. *Mind in Society: The Development of Higher Psychological Processes.* Cambridge MA: Harvard University Press, 1978.

Welsh Assembly Government/Welsh Enterprise Educators Network. *Feasible, Desirable or What? An Investigation into the Feasibility of an Entrepreneurship Educators' Programme in Wales.* Cardiff: WEEN, 2010.

Wertheimer, Michael. "Experimentelle Studien über das Sehen von Bewegung." *Zeitschrift für Psychologie* 61 (1912): 161–265.

Wilson, Tim. *A Review of Business University Collaboration.* Report commissioned by Secretary of State for Business Innovation and Skills, the Right Hon. Vince Cable MP, and the Minister for Science and Universities, the Right Hon. David Willetts MP. London, 2012.

World Economic Forum. *Insight Report: Enhancing Europe's Competitiveness: Fostering Innovation-driven Entrepreneurship in Europe.* World Economic Forum, 2014.

Zaltman, Gerald. *How Customers Think: Essential Insights into the Mind of the Market,* Boston: Harvard Business School Press, 2003

4. Entrepreneurial Universities in Egypt: Opportunities and Challenges

David A. Kirby and Nagwa Ibrahim

Introduction

Since the publication of Birch's seminal research on job generation in the United States,[1] governments around the world have viewed new and small businesses as the panacea for job generation and innovation. Hence, they have encouraged the creation of an enterprise culture and the promotion of entrepreneurship. This has been done in a variety of ways,[2] for example, by providing both "hard" support like finance and premises, and "soft" support such as advice, training, or consultancy. Perhaps unsurprisingly in a knowledge economy, most countries have tried to effect this through the education system. Initially, this was done through the provision of educational programs at the undergraduate and graduate levels,[3] but increasingly there is recognition that such a broad concept requires educational institutions themselves to be entrepreneurial. Accordingly, the concept of entrepreneurial universities has emerged in recent years, and much has been written about them: what they are, how they are created, and how they are managed.[4] In Egypt, the concepts of entrepreneurial universities and entrepreneurial education at the undergraduate and graduate levels are relatively new and neither appears to be highly developed.[5] Ashraf Sheta has observed the challenges when commenting on the proposal for an entrepreneurship curriculum in Egypt:

When commenting on the proposal for an entrepreneurial curriculum in Egypt, Sheta observed that several challenges residing in the educational system will need to be tackled, namely: the current pedagogical approach to teaching and learning; the number of students; the prevailing misunderstanding about the role of the university in community development; and the gap between education outputs and market needs.[6]

This is undoubtedly correct, but it is also likely that even more challenges will be encountered, especially when attempting to create more entrepreneurial universities. At the same time, there are likely to be opportunities. It is the purpose of this chapter to consider the nature and scale of these challenges and opportunities and the support that universities need to address them.

Literature Review
Entrepreneurial universities
While the twenty-first century knowledge economy has shown considerable interest in entrepreneurial universities, there is no agreed definition for this term. Kirby et al.[7] have shown that numerous definitions exist for entrepreneurial universities. These range from "the entrepreneurial university is nothing more than a seller of services in the knowledge industry,"[8] to "the entrepreneurial university is a natural incubator, providing support structures for teachers and students to initiate new ventures: intellectual, commercial and conjoint."[9] Despite the lack of a standard definition, there are numerous examples and case studies of entrepreneurial universities.[10] Entrepreneurial universities do exist, and, as has been pointed out,[11] it would seem that there is a set of formal and informal institutional factors that can hinder or facilitate their development. One of the aims of this research is to identify those factors in the Egyptian context. If Egyptian universities are to play a role in promoting entrepreneurship in the country, they will need to address the barriers that are preventing entrepreneurship in their organizations. This will involve creating an environment/ecosystem that is supportive of and conducive to development, including clear and fair policies and procedures that are communicated positively and enthusiastically, as Birley has suggested in a different geographical context.[12]

Entrepreneurial education
Entrepreneurship has been taught in universities in the United States since the late 1940s, with the first documented course offered at Harvard Business School.[13] Elsewhere, such courses have only recently been

introduced. In the United Kingdom and Europe, for instance, the first university courses in entrepreneurship were not taught until the 1980s. As in the United States, they were intended to encourage students to become self-employed and start their own ventures after graduation. While this remains a key objective of many programs,[14] over the years an ongoing debate has developed not just over whether entrepreneurship can be taught[15] but also over the nature and purpose of entrepreneurship education. Is it about new venture creation and equipping students with the functional skills to start a business, or is it a much broader concept that recognizes entrepreneurship as a way of thinking and behaving, of seeing opportunities and harnessing the resources to bring such opportunities to fruition—and bringing about change in the process?[16] Traditionally, programs have focused on teaching students how to write a business plan,[17] but it is increasingly recognized that effective entrepreneurship education should engage students with the various thinking styles and behaviors associated with the entrepreneur and not just with a set of business tools.[18] Accordingly, traditional studies on the impact of entrepreneurship education have tended to focus on entrepreneurial intention and the number of new ventures created, often without reference to the program aims or pedagogy.[19] Indeed, a valid program objective may in fact be to deter those who are not suited for self-employment or who do not have innovative entrepreneurial ideas from embarking on this course of action. It might be equally valid not to create entrepreneurs but intrapreneurs—people who see opportunities and bring about change not by creating new ventures but by changing existing organizations.[20]

The starting point for understanding impact, therefore, must be understanding the program's objective: whether its purpose is to educate students *about* entrepreneurship (that is, raise their awareness of it) or to educate them *for* entrepreneurship (develop their entrepreneurial abilities and tendencies). The program's focus could be to equip students with the knowledge and skills to start a business after graduation, or to develop skills, abilities, attitudes, and patterns of behavior of an enterprising or entrepreneurial individual.[21]

This broader concept of educating students for entrepreneurship is increasingly being adopted. As the European Commission has recognized, "the benefits of entrepreneurship education are not limited to startups, innovative ventures and new jobs" but also encompass "an individual's ability to turn ideas into action and is therefore a key competence for all, helping young people to be more creative and self-confident in whatever they undertake."[22] This being the case, there needs to be a very significant transformation not just in what is taught but in the process and pedagogy of learning—which constitutes a paradigm shift.[23] Much has been written about the changes that are needed, and the purpose of this study is to determine what Egyptian universities understand entrepreneurship education to be, and to establish if and how they provide it, while at the same time examining the related challenges and opportunities.

The Egyptian Context

In Egypt, there is an increasing awareness of the need to promote entrepreneurship and develop an enterprise culture. Mamdouh has recognized this,[24] and the Egyptian National Competitiveness Council has pointed to the fact that Egypt lags behind other countries in terms of its capacity for innovation.[25] As elsewhere, education has a major role to play in promoting awareness and understanding of entrepreneurship and the need for innovation. The National Competitiveness Council report and the 2008 and 2010 Global Entrepreneurship Monitor (GEM) reports for Egypt recognize that education needs strengthening if Egypt is to compete successfully in the modern global economy. Indeed, in both of the GEM reports, national experts asserted that the Egyptian education system was very weak with respect to entrepreneurship, as the country was ranked last out of the fifty-three GEM countries for this field.[26] As a consequence, in July 2010, the Ministry of Higher Education initiated a project in partnership with the Middle East Council for Small Business and Entrepreneurship that was intended to introduce courses in entrepreneurship and small business management, logic, and critical thinking and innovation to the faculties of agriculture, commerce, economics, engineering, political science, and science in three of the leading state universities in the country, located in Alexandria, Cairo, and Helwan.[27] Prior to the January 2011 revolution, the endeavor was projected to run from 2011–15 and was to involve over 100,000 students and ninety professors before being rolled out to nineteen state universities in the form of specialist entrepreneurship tracks and graduate-level degrees. This ambitious proposal was backed by extensive international experience through the academic network of the International Council for Small Business. Yet the justification for the program is unclear, as are the program's pedagogic processes and precise aims and objectives. Even so, it is possible to agree with Sheta that "entrepreneurship education in Egypt is a necessity for moving from a factor-driven to an innovation-driven economy, and also to enhance competitiveness at the macro-level, leading to the bottom line effect of improving living conditions."[28]

Purpose of the Study

Against this academic and contextual background, the purpose of this chapter is the following:

1. To identify:
   a. What Egyptian universities mean by the term "entrepreneurial universities."
   b. The formal and informal institutional factors that potentially hinder or facilitate the development of entrepreneurial universities in Egypt.
   c. The plans Egyptian universities have for their institutions in relation to entrepreneurship.
2. To determine what Egyptian universities understand entrepreneurship education to be.
3. To establish if and how Egyptian universities provide entrepreneurial education.
4. To examine the challenges and opportunities Egyptian universities face.

Methodology

To achieve these objectives, a detailed questionnaire survey was undertaken at five Egyptian universities, including three state universities, one private international university, and one private Egyptian university. Only one of the three state universities (which averaged 144,667 students) was part of the aforementioned entrepreneurship pilot project.[29] The private universities were considerably smaller, averaging only 4,912 students. We refer to each university as a respondent in the findings below.

We acknowledge that the responses are not necessarily representative of higher education in Egypt as a whole, as the country has 29 state-funded and 39 private universities. They do, however, provide indications of the opportunities and challenges facing the sector as it strives to become more entrepreneurial, to contribute to the creation of an entrepreneurial culture/ecosystem, and to train enterprising graduates.

The questionnaire, which was in English, was divided into six sections:

A. Details of the institution and its approach to entrepreneurship.
B. Institutional attitudes toward entrepreneurship education.
C. Details of each entrepreneurship program in the institution.

D. The institution's future plans.

E. The difficulties/challenges the institution has experienced and/or expects to experience.

F. The support needed to enable the institution to become more entrepreneurial and to develop entrepreneurship education.

Survey questions were based on evidence from the literature and comprised a series of open and closed questions, providing respondents with an opportunity to express their opinions in their own words.

Findings: Opportunities and Challenges
Institutional approach to entrepreneurship

The respondents were presented with five definitions of entrepreneurial universities. These defined an entrepreneurial university as one that:

- Involves the creation of new business ventures by university professors, technicians, or students.[30]
- Seeks to innovate in how it goes to business. It seeks to work out a substantial shift in organizational character.[31]
- Is a natural incubator, providing support structures for teachers and students to initiate new ventures.[32]
- Sells services in the knowledge industry.[33]
- Has the ability to innovate, recognize and create opportunities, take risks, and respond to challenges.[34]

Respondents from the five universities supported the definitions by Chrisman, Etzkowitz, Williams, and Kirby, suggesting that an acceptable definition of an entrepreneurial university is one that "has the ability to innovate, recognize and create opportunities, take risks, and respond to challenges. It sells its services in the knowledge industry and is a natural incubator that supports its academics, technicians, and students to create new ventures."[35]

However, having recognized the characteristics of an entrepreneurial university, only one of the responding institutions identified itself as such. The main obstacles were seen to be:

- The fact that entrepreneurship is not perceived as a primary function of universities.
- The organizational structures and governance of the universities, which restricted innovation and the opportunity for change.

Administrative clashes with research objectives and teaching objectives, traditional ways of teaching, and a lack of experience were negative factors cited by four of the five respondents. Similarly, the respondents highlighted the lack of physical resources (accommodation), links with industry, an appropriate reward system, and adequate funding as critical issues preventing their universities from becoming entrepreneurial. The "lack of knowledge and experience" was seen by one university as the key factor preventing it from becoming entrepreneurial, while for another, it was the "lack of an entrepreneurial culture within university staff, the lack of pertinence for the university due to the economic conditions, and the mistrust between industry and the university."

Despite this, all five universities claimed to have responded to the challenge to become more entrepreneurial and had plans to continue to do so. None of the universities had introduced an entrepreneurship qualification at the undergraduate level, but four had introduced undergraduate-level entrepreneurship modules, and two had introduced either modules or a qualification in entrepreneurship at the postgraduate level, including an opportunity to study for a doctorate at one state university. While none of the institutions had introduced policies to encourage its academic staff to create new ventures, three had provided a service to advise both staff and students on how to create a new business, and two claimed to offer seed funding for both staff and students. One institution was providing accommodation for university spinout and startup businesses. In three institutions, academic entrepreneurship centers had been established. For two of those universities, the aim of the centers was to promote research in the field of entrepreneurship, while at the third, the center was intended to offer entrepreneurship courses.

In terms of their plans for further developments, four of the universities talked about opening a technology transfer office, but otherwise there was no consensus. For all the respondents, broad areas for planned future initiatives related to:

- Education: introducing a master's program, offering more advanced courses, and integrating entrepreneurship education into the core curriculum.
- Organization and governance: making entrepreneurship a primary function of the university, changing the organization structure, and developing an entrepreneurship center.
- Support: building an incubator and science park, providing seed funding, developing links with the business community, becoming part of a business cluster, and providing training.

Institutional attitudes toward entrepreneurship education

All five of the universities offered entrepreneurship programs—four within the curriculum, and one as an extracurricular program. Four of the respondents stated that their aim was either to create graduate entrepreneurs (that is, students who will start their own business after graduation) and/or to develop a set of transferable skills that students could use on leaving university. Only two universities believed that entrepreneurship education was about changing the way students think and behave, and only one institution claimed to have a consistent definition of entrepreneurship education. Another respondent said that its definition varied according to the program's learning objectives, while the remaining three universities said the definition varied from department to department. Two universities said their working definition of entrepreneurship education was determined by whether the course was in or outside the curriculum, while two others said it depended on the level of the course.

In total, the five universities offered nine courses that typically ran for one semester. All nine courses focused on teaching the participants about entrepreneurship, but five of these courses also intended to educate participants to *become* entrepreneurs. Only three universities expected their students to start a business upon completing the program, despite the supposed course aims. All nine courses were expected to have an impact on the students' attitudes and behavior, though this was only a stated objective of two of the universities.

Entrepreneurship programs

Although entrepreneurship is traditionally taught in business schools or departments, five of the nine programs were offered in non-business faculties, and often in the engineering department. This is an interesting development and something to be encouraged, especially if it results in cross-disciplinary collaboration and the breaking down of subject boundaries. Only three of the universities had compulsory entrepreneurship programs; the other two had optional ones, and only enrolled a relatively small numbers of students—usually no more than fifty. Only one compulsory program in one of the private universities had over one hundred students. All of the programs involved traditional lectures, while seven courses included talks by successful entrepreneurs, and five offered talks by startup entrepreneurs, as well as practicing bankers, business angels, venture capitalists, marketing experts, accountants, and other professionals. These offerings were the strongest at the two private universities. In many cases, there is evidence of experiential learning through case studies (six courses), simulations (six courses), writing a startup assignment/project (five courses), or preparing a business plan (four courses). In only two cases were students required to undertake a placement or internship with a small or medium enterprise.

Overall, there is evidence of the sort of teaching that is required for entrepreneurial education, but there is little evidence of the practical support necessary to facilitate both real-life experiential learning and the establishment of new ventures. None of the institutions had a science park, only two provided startup accommodation and seed funding, and only three offered ongoing mentoring to startup ventures.

Difficulties and challenges

When asked about existing or expected challenges, the respondents primarily referred to a lack of resources. A shortage of time and funding was also mentioned, but the main concern was the lack of understanding and awareness of the need for entrepreneurship in universities, coupled with a lack of expertise not just to teach the subject but also to help commercialize new ideas and intellectual property. Reference was also made to the lack of trust in business, especially in the state sector. Many professors believe that business and commerce is "a fifth column gnawing away at the basic values that define a university."[36] Hence, there will be resistance to introducing new entrepreneurial courses. This needs to be overcome. Egypt's universities need to learn to work more closely with the business community, and vice versa.

Resistance can also arise from what can be termed "turfiness."[37] Traditionally, entrepreneurship has been taught in business schools. At least one respondent, who was not affiliated with a business school, acknowledged that attempts to introduce entrepreneurship to the curriculum had been opposed by business school colleagues elsewhere in the university. The opposition was overcome in this particular case, but this issue will doubtless continue to arise. While understandable, it has to be resisted. Entrepreneurship is for everyone, not just for business students.[38]

Bringing about change is hampered by bureaucracy and Egypt's perceived risk-averse culture, which makes it difficult for universities to innovate. Egyptian universities are not unique in this respect. Universities have traditionally been slow to change, as they exhibit many of the inherent barriers to transformation possessed by large organizations.[39] In particular, their hierarchical structures, which require many levels of approval, and the need for control via adherence to rules and procedures work against change, as does their organization into academic silos, or departments. Several respondents pointed to the need for internal reorganization to encourage greater flexibility, while one suggested a lessening of control over the curriculum by the Egyptian Government's Supreme Council.

In addition, like their international counterparts, Egyptian universities have a set of traditional academic values. When confronted with these new developments, many professors believe that such measures "will drive out their other more fundamental university qualities, such as: intellectual integrity, critical inquiry, and commitment to learning and understanding."[40] But, as Etzkowitz has demonstrated,[41] some of the most successful universities from an entrepreneurial perspective are among the world's leading research-led academic institutions. Indeed, this was a further challenge recognized by at least one of the respondents: how to reorient the university's research so that it better matched the needs of the market.

Support needed

When asked what support was needed to help promote entrepreneurship and entrepreneurship education in universities, the survey respondents made a variety of suggestions. The list was topped by a need for greater funding and resources, but reference was also made to the need for more appropriate reward systems, training and staff development, reduced bureaucracy, and greater autonomy. In particular, respondents called for capacity building so that more staff understand the concept of entrepreneurship and are able to teach it; for more appropriate physical resources, such as enterprise laboratories, hatcheries, incubators, and science parks; and for the creation of practitioner networks including entrepreneurs, investors, and legal experts. These aspects can be developed by the institutions themselves, but as several of the respondents acknowledged, this requires that entrepreneurship be recognized as a primary function of the university. In order to effect this understanding, Egypt's universities "will need to formulate a high-level strategy that demonstrates the university's intent, makes it clear that the university encourages this form of behavior, provides the university's staff with the knowledge and support to start their own businesses, and creates an environment that reduces the risk."[42]

Conclusion

The results of the research indicate that respondents clearly understand what an entrepreneurial university is, with none (save one private institution) believing that their university met the requisite criteria. The reasons they give for this lend support to the findings of Kirby et al.,[43] which indicate that the main barriers to a university becoming more entrepreneurial are its organizational structure and governance, the belief that entrepreneurship is not a primary function of a university, inadequate links with industry, the belief that entrepreneurship conflicts with the university's research objectives, a lack of entrepreneurial experience, and insufficient funding and physical resources.

With respect to entrepreneurship education, all five of the universities surveyed offer programs in entrepreneurship, though only one of them was part of the pilot project offered by the Ministry of Higher Education and the Middle East Council for Small Business and Entrepreneurship.[44] While the responses suggest that the universities are adopting an innovative and experiential approach to learning, the results mask very considerable differences in delivery between the three state universities and the two private universities. None of the institutions provided hatcheries or incubators in which students could launch their ventures as part of the learning process, and only one required its students to undertake an internship in a small or medium enterprise. Subsequent to the study, though, four universities went on to establish technology transfer offices using European Union funding. These increase industry-academia collaboration, support the transfer of technology from universities to market, and advise university staff on patenting and spinoff activities.

These results have considerable implications for the universities themselves, and for Egypt at large. First, the universities need to be permitted to be more autonomous and to be able to organize themselves so as to be more flexible and innovative. Second, bureaucracy needs to be reduced both within the institutions themselves and between the institutions and the state. Third, universities and the business community need to be encouraged to establish closer links, and the country's universities need to be perceived as part of what Leydesdorff and Etzkowitz have termed an entrepreneurship-fostering "Triple Helix" network of government, the university sector, and the industry/business sector.[45] This can be facilitated, at least in part, through the creation of technology transfer offices, science parks, and incubators. Several of the respondents planned to develop these institutions, and they have already begun to emerge, with the American University in Cairo's Venture Lab being a notable example. Fourth, there needs to be a capacity-building program whereby academic and administrative university staff can learn to appreciate why entrepreneurship is so important, and what is needed to deliver it.

Perhaps most important, and somewhat paradoxically, a national policy is needed that shifts the university paradigm "from purely upholding the mission of research and teaching to the . . . mission of promoting economic and social development."[46] This would make it clear that entrepreneurship is a primary objective of all higher education institutions in the modern Egyptian economy. Education reform will "not work without policies in place to change the existing metrics."[47] Further, in accordance with Ajzen's theory of planned behavior,[48] people will respond if they believe it is a social norm or expectation. The Egyptian government, therefore, needs to have a clear vision of the role its universities should play, and a strategy for bringing about the educational reforms that are required if the country is to compete effectively in the modern global economy.[49] As elsewhere, though, the state needs to "steer from a distance,"[50] and Egyptian universities must move away from close government regulation and sector standardization and search for their own special organizational identities, as Clark has argued.[51]

While based on a very small sample of Egyptian state and private universities, the findings of the present study corroborate those of Kirby et al.,[52] and would seem to fit with earlier theoretical considerations of the topic based on Egypt,[53] as well as developments elsewhere.[54] While the country has begun to recognize the importance of entrepreneurship and the role that universities play in developing an entrepreneurship ecosystem, the study shows that much more needs to be done by both the universities themselves and by the state. A triple helix of Universities–Industry–Government needs to be created, and education in general, and higher education in particular, needs to be recognized in policies to promote entrepreneurship in the country, something that has not happened.[55] The evidence suggests that this is beginning to happen but, as elsewhere, it will take time to effect the very necessary changes that are needed.

Notes

1   David Birch, *The Job Generation Process* (Massachusetts: MIT Program on Neighborhood and Regional Change, 1979); David Birch, *Job Creation in America: How Our Smallest Companies Put the Most People to Work* (New York: Free Press, 1987).
2   Anders Lundström and Lois A. Stevenson, *Entrepreneurship Policy: Theory and Practice* (New York: Springer, 2005).
3   Jerome A. Katz, "The Chronology and Intellectual Trajectory of American Entrepreneurship Education, 1876–1999," *Journal of Business Venturing* 18 (2003): 283–300; Donald F. Kuratko, "The Emergence of Entrepreneurship Education: Development, Trends and Challenges," *Entrepreneurship Theory and Practice* 29 (2005): 577–97; George Solomon et al., "The State of

Entrepreneurship Education in the United States: A Nationwide Survey and Analysis," *International Journal of Entrepreneurship Education* 1 (2002): 65–86; Karl Vesper and William Gartner, *University Entrepreneurship Programs Worldwide* (Los Angeles, CA: The University of South Carolina, 1998); Todd A. Finkle and David Deeds, "Trends in the Market for Entrepreneurship Faculty during the Period 1989–1998," *Journal of Business Venturing* 16 (2001): 613–30.

4   David A. Kirby, David Urbano and Maribel Guerro, "Making Universities More Entrepreneurial: Development of a Model," *Canadian Journal of Administrative Sciences* 28 (2011): 303–16.

5   David A. Kirby and Nagwa Ibrahim, "An Enterprise Revolution for Egyptian Universities," *Education, Business and Society: Contemporary Middle Eastern Issues* 8 (2012): 98–111; Ashraf Sheta, "Developing an Entrepreneurship Curriculum in Egypt: The Road Ahead" (paper presented at the ICSB World Conference, Stockholm, Sweden, June 15–18, 2011).

6   Sheta, "Developing an Entrepreneurship Curriculum," 4–5.

7   Kirby, Urbano, and Guerro, "Making Universities More Entrepreneurial."

8   Gareth Williams, *The Enterprising University: Reform, Excellence and Equity* (Buckingham: The Society for Research into Higher Education and Open University Press, 2003), 14.

9   Henry Etzkowitz, "Research Groups as 'Quasi Firms': The Invention of the Entrepreneurial University," *Research Policy* 32 (2003): 112.

10   Andrés Bernasconi, "University Entrepreneurship in a Developing Country: The Case of P. Universidad Catolica de Chile, 1985–2000," *Higher Education* 50 (2005): 247–74; Burton R. Clark, *Creating Entrepreneurial Universities: Organisational Pathways of Transformation* (Oxford: Elsevier Science, 1998); Deanna De Zilwa, "Using Entrepreneurial Activities as a Means of Survival: Investigating the Processes Used by Australian Universities to Diversify Their Revenue Streams," *Higher Education* 50 (2005): 387–411; Merle Jacob, Mats Lundquist, and Hans Hellsmark, "Entrepreneurial Transformations in the Swedish University System: The Case of Chalmers University of Technology," *Research Policy* 32 (2003): 1555–69; David A. Kirby, "Creating Entrepreneurial Universities in the UK: Applying Entrepreneurial Theory to Practice," *Journal of Technology Transfer* 31 (2006a): 599–603; Rory P. O'Shea, Thomas J. Allen, Kenneth P. Morse, Colm O'Gorman, and Frank Roche, "Delineating the Anatomy of an Entrepreneurial University: The Massachusetts Institute of Technology Experience," *R&D Management* 37 (2007): 1–16; Liana M. Ranga, Koenraad Debackere, and Nick Von-Tunzelmann, "Entrepreneurial Universities and the Dynamics of Academic Knowledge Production: A Case Study of Basic vs. Applied Research in Belgium," *Scientometrics* 58 (2003): 301–20; Robert Tijssen, "Universities and Industrially Relevant Science: Towards Measurement Models and Indicators of Entrepreneurial Orientation," *Research Policy* 35 (2007): 1569–85; Keiko Yokoyama, "Entrepreneurialism in Japanese and UK Universities: Governance, Management, Leadership and Funding," *Higher Education* 52 (2006): 523–55; Fang Zhao, "Academic Entrepreneurship: Case Study of Australian Universities," *International Journal of Entrepreneurship and Innovation* 5 (2004): 91–97.

11   Kirby, Urbano and Guerro, "Making Universities More Entrepreneurial."

12   Sue Birley, "Universities, Academics and Spinout Companies: Lessons from Imperial," *International Journal of Entrepreneurship Education* 1 (2002): 133–53.

13 Robert H. Brockhaus et al., *Entrepreneurship Education: A Global View* (Aldershot: Ashgate Publishing, 2001).

14 Abdulhasan Al-Dairi, Ronald McQuaid, and John Adams, "Entrepreneurship Training to Promote Start-ups and Innovation in Bahrain," *International Journal of Innovation and Knowledge Management in the Middle East and North Africa* 1 (2012): 179–210.

15 Colette Henry, Frances Hill, and Claire Leitch, "Entrepreneurship Education and Training: Can Entrepreneurship Be Taught? Part II," *Education + Training* 47 (2005): 158–69.

16 David A. Kirby, "Entrepreneurship Education: Can Business Schools Meet the Challenge?" in *International Entrepreneurship Education: Issues and Newness*, ed. Alain Fayolle and Heinz Klandt (Cheltenham: Edward Elgar, 2006), 35–54; Alain Fayolle and Benoît Gailly, "From Craft to Science: Teaching Models and Learning Process in Entrepreneurship Education," *Journal of European Industrial Training* 32 (2008): 569–93.

17 Lorella Cannavacciuolo et al., "To Support the Emergence of Academic Entrepreneurs: The Role of Business Plan Competitions," in *International Entrepreneurship Education: Issues and Newness*, ed. Alain Fayolle and Heinz Kland (Cheltenham: Edward Elgar, 2006), 55–73; Benoît Gailly, "Can You Teach Entrepreneurs to Write Their Business Plan? An Empirical Evaluation of Business Plan Competitions," in *International Entrepreneurship Education: Issues and Newness*, ed. Alain Fayolle and Heinz Klandt (Cheltenham: Edward Elgar, 2006), 133–54.

18 David A. Kirby, *Entrepreneurship* (Maidenhead, UK: McGraw-Hill, 2003).

19 Georg von Graevenitz, Dietmar Harhoff, and Richard Weber, "The Effects of Entrepreneurship Education," *Journal of Economic Behavior and Organization* 76 (2010): 90–112; Lars Kolvereid and Øystein Moen, "Entrepreneurship among Business Graduates: Does a Major in Entrepreneurship Make a Difference?" *Journal of European Industrial Training* 21 (1997): 154; Etienne R. Mentoor and Chris Friedrich, "Is Entrepreneurial Education at a South African University Successful? An Empirical Example," *Industry and Higher Education* 21 (2007): 221–32; Nicole E. Peterman and Jessica Kennedy, "Enterprise Education: Influencing Students' Perceptions of Entrepreneurship," *Entrepreneurship Theory and Practice* 28 (2003): 129–44; Vangelis Souitaris, Stefania Zerbaniti, and Andreas Al-Laham, "Do Entrepreneurship Programmes Raise Entrepreneurial Intention of Science and Engineering Students? The Effect of Learning, Inspiration and Resources," *Journal of Business Venturing* 22 (2007): 566–91.

20 David A. Kirby and Harim Humayun, "Entrepreneurship Education and the Attitudes and Intentions of Students: An Empirical Study in Egypt," *International Journal of Management* 30 (2013): 23–35.

21 Ghulam Nabi and Rick Holden, "Graduate Entrepreneurship: Intentions, Education and Training," *Education + Training* 50 (2008): 545–51.

22 European Commission, *Entrepreneurship in Higher Education, Especially within Non-Business Studies: Final Report* (Brussels: European Commission, 2008), 7.

23 David A. Kirby, "Changing the Entrepreneurial Education Paradigm," in *Handbook of Research in Entrepreneurship Education. Volume 1: A General Perspective*, ed. Alain Fayolle (Cheltenham: Edward Elgar, 2007), 21–45.

24 Farid Mamdouh, "Entrepreneurship in Egypt and the US Compared: Directions for Further Research Suggested," *Journal of Management Development* 26 (2007): 428–40.

25   Egyptian National Competitiveness Council, *The Egyptian Competitiveness Report: Towards a Competitive Egypt Where Everybody Wins* (Cairo: Egyptian National Competitiveness Council, 2008).

26   Hala Hattab, *Global Entrepreneurship Monitor: Egypt Entrepreneurship Report, 2008* (Cairo: The Industrial Modernization Centre, 2009); Hala Hattab, *Global Entrepreneurship Monitor: Egypt Entrepreneurship Report, 2010* (Cairo: The Industrial Modernization Centre, 2012).

27   Sheta, "Developing an Entrepreneurship Curriculum."

28   Sheta, "Developing an Entrepreneurship Curriculum," 9.

29   Sheta, "Developing an Entrepreneurship Curriculum."

30   James Chrisman, Timothy Hynes, and Shelby Fraser, "Faculty Entrepreneurship and Economic Development: The Case of the University of Calgary," *Journal of Business Venturing* 10 (1995): 267–81.

31   Clark, *Creating Entrepreneurial Universities.*

32   Etzkowitz, "Research Groups as 'Quasi Firms.'"

33   Williams, *The Enterprising University.*

34   Kirby, "Entrepreneurial Universities."

35   Sheta, "Developing an Entrepreneurship Curriculum."

36   Ian McNay, "The E-factors and Organisation Cultures in British Universities," in *The Enterprising University: Reform, Excellence and Equity*, ed. Gareth Williams (Buckingham: The Society for Research into Higher Education and Open University Press, 2003), 20–28.

37   Gifford Pinchot, *Intrapreneuring* (New York: Harper and Row, 1985).

38   European Commission, *Entrepreneurship in Higher Education.*

39   Kirby and Ibrahim, "Enterprise Revolution."

40   Williams, *The Enterprising University*, 19.

41   Etzkowitz, "Research Groups as 'Quasi Firms.'"

42   Kirby and Ibrahim, "Enterprise Revolution."

43   Kirby et al., "Making Universities More Entrepreneurial."

44   Sheta, "Developing an Entrepreneurship Curriculum."

45   Loet Leydesdorff and Henry Etzkowitz, "The Transformation of University–Industry–Government Relations," *Electronic Journal of Sociology* 5 (2001): 1–17.

46   Ka Ho Mok, "Fostering Entrepreneurship: Changing Role of Government and Higher Education Governance in Hong Kong," *Research Policy* 34 (2005): 560.

47   World Economic Forum, *Global Education Initiative: MENA Roundtable on Entrepreneurship Education* (Geneva: World Economic Forum, 2010), 22.

48   Icek Ajzen, "The Theory of Planned Behavior," *Organizational Behavior and Human Decision Processes* 50 (1991): 179–211.

49   Egyptian National Competitiveness Council, *The Egyptian Competitiveness Report: Towards a Competitive Egypt Where Everybody Wins* (Cairo: Egyptian National Competitiveness Council, 2008).

50   Mok, "Fostering Entrepreneurship."

51   Clark, *Creating Entrepreneurial Universities.*

52   Kirby et al., "Making Universities More Entrepreneurial."

53   Kirby and Ibrahim, "Enterprise Revolution."

54   Mok, "Fostering Entrepreneurship."

55   Kirby and Ibrahim, 2013.

Bibliography

Ajzen, Icek. "The Theory of Planned Behavior." *Organizational Behavior and Human Decision Processes* 50 (1991): 179–211.

Bernasconi, Andrés. "University Entrepreneurship in a Developing Country: The Case of P. Universidad Catolica de Chile, 1985–2000." *Higher Education* 50 (2005): 247–74.

Birch, David. *Job Creation in America: How Our Smallest Companies Put the Most People to Work*. New York: Free Press, 1987.

———. *The Job Generation Process*. Massachusetts: MIT Program on Neighborhood and Regional Change, 1979.

Birley, Sue. "Universities, Academics and Spinout Companies: Lessons from Imperial." *International Journal of Entrepreneurship Education* 1 (2002): 133–53.

Brockhaus, Robert, Gerald E. Hills, Heinz Klandt, and Harold P. Welsch. *Entrepreneurship Education: A Global View*. Aldershot: Ashgate Publishing, 2001.

Cannavacciuolo, Lorella, Guido Capaldo, Gianluca Esposito, Luca Iandoli, and Mario Raffa. "To Support the Emergence of Academic Entrepreneurs: The Role of Business Plan Competitions." In *International Entrepreneurship Education: Issues and Newness*, edited by Alain Fayolle and Heinz Klandt, 55–73. Cheltenham: Edward Elgar, 2006.

Chrisman, James, Timothy Hynes, and Shelby Fraser. "Faculty Entrepreneurship and Economic Development: The Case of the University of Calgary." *Journal of Business Venturing* 10 (1995): 267–81.

Clark, Burton R. *Creating Entrepreneurial Universities: Organisational Pathways of Transformation*. Oxford: Elsevier Science, 1998.

Al-Dairi, Abdulhasan, Ronald McQuaid, and John Adams. "Entrepreneurship Training to Promote Startups and Innovation in Bahrain." *International Journal of Innovation and Knowledge Management in Middle East and North Africa* 1 (2012): 179–210.

De Zilwa, Deanna. "Using Entrepreneurial Activities as a Means of Survival: Investigating the Processes Used by Australian Universities to Diversify Their Revenue Streams." *Higher Education* 50 (2005): 387–411.

Egyptian National Competitiveness Council. *The Egyptian Competitiveness Report: Towards a Competitive Egypt Where Everybody Wins*. Cairo: Egyptian National Competitiveness Council, 2008.

Etzkowitz, Henry. "Research Groups as 'Quasi Firms': The Invention of the Entrepreneurial University." *Research Policy* 32 (2003): 109–21.

European Commission. *Entrepreneurship in Higher Education, Especially within Non-Business Studies: Final Report*. Brussels: European Commission, 2008.

Fayolle, Alain, and Benoît Gailly. "From Craft to Science: Teaching Models and Learning Process in Entrepreneurship Education." *Journal of European Industrial Training* 32 (2008): 569–93.

Finkle, Todd A., and David Deeds. "Trends in the Market for Entrepreneurship Faculty during the Period 1989–1998." *Journal of Business Venturing* 16 (2001): 613–30.

Gailly, Benoît. "Can You Teach Entrepreneurs to Write Their Business Plan? An Empirical Evaluation of Business Plan Competitions." In *International Entrepreneurship Education: Issues and Newness*, edited by Alain Fayolle and Heinz Klandt, 133–54. Cheltenham: Edward Elgar, 2006.

Graevenitz, Georg von, Dietmar Harhoff, and Richard Weber. "The Effects of Entrepreneurship Education." *Journal of Economic Behavior and Organization* 76 (2010): 90–112.

Hattab, Hala. *Global Entrepreneurship Monitor: Egypt Entrepreneurship Report, 2008.* Cairo: The Industrial Modernization Centre, 2009.

———. *Global Entrepreneurship Monitor: Egypt Entrepreneurship Report, 2010.* Cairo: The Industrial Modernization Centre, 2012.

Henry, Colette, Frances Hill, and Claire Leitch. "Entrepreneurship Education and Training: Can Entrepreneurship Be Taught? Part II." *Education + Training* 47 (2005): 158–69.

Jacob, Merle, Mats Lundquist, and Hans Hellsmark. "Entrepreneurial Transformations in the Swedish University System: The Case of Chalmers University of Technology." *Research Policy* 32 (2003): 1555–69.

Katz, Jerome A. "The Chronology and Intellectual Trajectory of American Entrepreneurship Education, 1876–1999." *Journal of Business Venturing* 18 (2003): 283–300.

Khaled, Ashraf. "Egypt: Uncertain Future for E-university." *University World News*, July 6, 2008. http://www.universityworldnews.com/article. php?story=20080703152831554

Kirby, David A. "Changing the Entrepreneurial Education Paradigm." In *Handbook of Research in Entrepreneurship Education. Volume 1: A General Perspective*, ed. A. Fayolle. Cheltenham: Edward Elgar, 2007.

———. "Creating Entrepreneurial Universities in the UK: Applying Entrepreneurial Theory to Practice." *Journal of Technology Transfer* 31 (2006): 599–603.

———. *Entrepreneurship.* Maidenhead, UK: McGraw-Hill, 2003.

———. "Entrepreneurship Education: Can Business Schools Meet the Challenge?" In *International Entrepreneurship Education: Issues and Newness*, edited by Alain Fayolle and Heinz Klandt, 35–54. Cheltenham: Edward Elgar, 2006.

Kirby, David A., and Harim. Humayun. "Entrepreneurship Education and the Attitudes and Intentions of Students: An Empirical Study in Egypt." *International Journal of Management* 30 (2013): 23–35.

Kirby, David A., and Nagwa Ibrahim. "An Enterprise Revolution for Egyptian Universities." *Education, Business and Society: Contemporary Middle Eastern Issues* 8 (2012): 98–111.

Kirby, David A., and Nagwa Ibrahim. "Entrepreneurship Education Policies in the MENA Region: Challenges and Opportunities." *American Journal of Entrepreneurship* 6 (2013), 1-15.

Kirby, David A., David Urbano, and Maribel Guerro. "Making Universities More Entrepreneurial: Development of a Model." *Canadian Journal of Administrative Sciences* 28 (2011): 303–16.

Kolvereid, Lars, and Øystein Moen. "Entrepreneurship among Business Graduates: Does a Major in Entrepreneurship Make a Difference?" *Journal of European Industrial Training* 21 (1997): 154–60.

Kuratko, Donald F. "The Emergence of Entrepreneurship Education: Development, Trends and Challenges." *Entrepreneurship Theory and Practice* 29 (2005): 577–97.

Leydesdorff, Loet, and Henry Etzkowitz. "The Transformation of University–Industry–Government relations." *Electronic Journal of Sociology* 5 (2001): 1–17.

Lundström, Anders, and Lois A. Stevenson. *Entrepreneurship Policy: Theory and Practice.* New York: Springer, 2005.

Mamdouh, Farid. "Entrepreneurship in Egypt and the US Compared: Directions for Further Research Suggested." *Journal of Management Development* 26 (2007): 428–40.

McNay, Ian. "The E-factors and Organisation Cultures in British Universities." In *The Enterprising University: Reform, Excellence and Equity*, ed. Gareth Williams, 20–28. Buckingham: The Society for Research into Higher Education and Open University Press, 2003.

Mentoor, Etienne R., and Chris Friedrich. "Is Entrepreneurial Education at a South African University Successful? An Empirical Example." *Industry and Higher Education* 21 (2007): 221–32.

Mok, Ka Ho. "Fostering Entrepreneurship: Changing Role of Government and Higher Education Governance in Hong Kong." *Research Policy* 34 (2005): 537–54.

Nabi, Ghulam, and Rick Holden. "Graduate Entrepreneurship: Intentions, Education and Training." *Education + Training* 50 (2008): 545–51.

O'Shea, Rory P., Thomas J. Allen, Kenneth P. Morse, Colm O'Gorman, and Frank Roche. "Delineating the Anatomy of an Entrepreneurial University: The Massachusetts Institute of Technology Experience." *R&D Management* 37 (2007): 1–16.

Peterman, Nicole E., and Jessica Kennedy. "Enterprise Education: Influencing Students' Perceptions of Entrepreneurship." *Entrepreneurship Theory and Practice* 28 (2003): 129–44.

Pinchot, Gifford. *Intrapreneuring*. New York: Harper and Row, 1985.

Ranga, Liana M., Koenraad Debackere, and Nick Von-Tunzelmann. "Entrepreneurial Universities and the Dynamics of Academic Knowledge Production: A Case Study of Basic vs. Applied Research in Belgium." *Scientometrics* 58 (2003): 301–20.

Sheta, Ashraf. "Developing an Entrepreneurship Curriculum in Egypt: The Road Ahead." Paper presented at the ICSB World Conference, Stockholm, Sweden, June 15–18, 2011.

Solomon, George, Susan Duffy, and Ayman El Tarabishi. "The State of Entrepreneurship Education in the United States: A Nationwide Survey and Analysis." *International Journal of Entrepreneurship Education* 1 (2002): 65–86.

Souitaris, Vangelis, Stefania Zerbaniti, and Andreas Al-Laham. "Do Entrepreneurship Programmes Raise Entrepreneurial Intention of Science and Engineering Students? The Effect of Learning, Inspiration and Resources." *Journal of Business Venturing* 22 (2007): 566–91.

Tijssen, Robert. "Universities and Industrially Relevant Science: Towards Measurement Models and Indicators of Entrepreneurial Orientation." *Research Policy* 35 (2007): 1569–85.

Vesper, Karl, and William Gartner. *University Entrepreneurship Programs Worldwide*. Los Angeles, CA: The University of South Carolina, 1998.

Von Graevenitz, Georg, Dietmar Harhoff, and Richard Weber. "The Effects of Entrepreneurship Education." *Journal of Economic Behavior and Organization* 76 (2010): 90–112.

Williams, Gareth. *The Enterprising University: Reform, Excellence and Equity*. Buckingham: The Society for Research into Higher Education and Open University Press, 2003.

World Economic Forum. *Global Education Initiative: MENA Roundtable on Entrepreneurship Education*. Geneva: World Economic Forum, 2010.

Yokoyama, Keiko. "Entrepreneurialism in Japanese and UK Universities: Governance, Management, Leadership and Funding." *Higher Education* 52 (2006): 523–55.

Zhao, Fang. "Academic Entrepreneurship: Case Study of Australian Universities." *International Journal of Entrepreneurship and Innovation* 5 (2004): 91–97.

5. Varieties of Entrepreneurs: The Entrepreneurship Landscape in Egypt

Ayman Ismail and Sherif Yehia

Introduction

Entrepreneurship has been a mainstay of economic research, literature, and policies in recent years. In this chapter, we adopt a broad and common definition of entrepreneurship as "any attempt at new business or new venture creation, such as self-employment, a new business organization, or the expansion of an existing business, by an individual, a team of individuals, or an established business."[1]

Our definition of entrepreneurship includes high-growth, innovation-driven enterprises, which are often labeled "opportunity entrepreneurs," as well as microenterprises and small and medium enterprises (SMEs), which are primarily driven by "necessity entrepreneurs." While these types of entrepreneurs are pillars of both advanced and developing economies, their impact varies. High-growth, innovation-driven entrepreneurship is a primary engine for economic growth, competitiveness, and renewal; small and medium enterprises are especially important for job creation and economic resilience; microenterprises contribute primarily to poverty alleviation; and social enterprises focus primarily on creating social impact.

Egypt is dominated by micro and small enterprises, most of which could be classified as necessity entrepreneurs, whom we define as those whose entrepreneurial actions arise out of a need, such as income generation. A growing number of support services have recently popped up to support the opportunity entrepreneur. Such services include financial, incubator, and governmental services.

In this chapter, we examine the entrepreneurship landscape in Egypt, covering the three main varieties of entrepreneurs, their motivations, impact, and how they fare in the country. In the second half of the chapter, we describe the current entrepreneurship ecosystem in Egypt. We examine important factors for entrepreneurship, such as access to finance, education and training, support services, and regulatory frameworks in order to determine the overall preparedness within Egypt to support the establishment and growth of entrepreneurs.

Varieties of Entrepreneurs and the Landscape in Egypt— Literature Review

'Opportunity' entrepreneurs versus 'necessity' entrepreneurs

The Global Entrepreneurship Monitor (GEM) defines a necessity entrepreneur as one who has "no other work options and needs a source of income;" in other words, one who is pushed into starting a business because they have no other, better options. An opportunity entrepreneur, on the other hand, tends to see the prospect of an opportunity and experiences more of a 'pull' than a 'push,' sometimes also seeking greater independence in one's work and/or an improvement in income.[2]

On the macro level, the World Economic Forum's *Global Competitiveness Report*[3] identifies three types of economies based on GDP per capita and share of exports comprising primary goods: factor-driven (generally characterized by subsistence and unskilled labor); efficiency-driven (more developed competitiveness and industrialization); and innovation-driven (knowledge-intensive, with a premium on service). Innovation-driven economies with higher GDP per capita tend to have higher rates of opportunity-driven entrepreneurship, while economies with lower GDP per capita tend to have higher rates of necessity entrepreneurship.[4]

In a low-opportunity and necessity-driven entrepreneurship economy, the industry structure is largely dominated by micro and informal enterprises. In an opportunity-driven entrepreneurship economy, the industry structure is dominated by strong small and medium sized enterprises in addition to large firms.[5] Egypt tends to be dominated by necessity-driven entrepreneurship, as illustrated by figure 5.1. A primarily factor-driven economy, Egypt has a slightly higher rate of necessity entrepreneurship than the average of other factor-driven

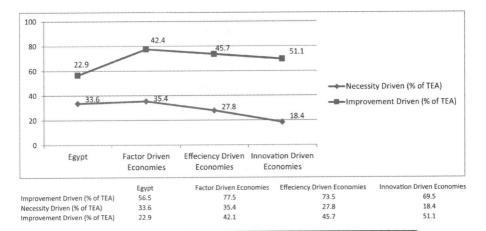

	Egypt	Factor Driven Economies	Effeciency Driven Economies	Innovation Driven Economies
Improvement Driven (% of TEA)	56.5	77.5	73.5	69.5
Necessity Driven (% of TEA)	33.6	35.4	27.8	18.4
Improvement Driven (% of TEA)	22.9	42.1	45.7	51.1

Figure 5.1 Necessity-driven vs. opportunity-driven TEA

countries, at 33 percent of total entrepreneurial activity (TEA), while its opportunity-driven entrepreneurship ranks lower than the average of other factor-driven countries.

Micro, small, and medium enterprises (MSMEs)

Much of the traditional entrepreneurship literature stemming from both Europe and the United States discusses the importance of small and medium enterprises (SMEs) for economic growth and job creation.[6] Policymakers around the world have responded to this research by fostering and promoting the SME landscape. Favorable tax policies and business-support services have been instituted in many countries to favor the creation and growth of the SME sector.

Defining the size of micro, small, or medium enterprises (MSMEs) is usually country-specific, depending on the size of the economy and level of development, and involves the use of variables such as number of employees, revenue or turnover, and company assets.[7] The International Finance Corporation (IFC) defines MSMEs as follows: microenterprises have 1–9 employees; small is 10–49 employees; and medium is 50–249 employees.[8]

In Egypt, the Central Bank of Egypt (CBE) and the Egyptian Banking Institute have defined small and very small enterprises as generating between LE100,000 and 10 million in annual revenue (USD14,000 to USD1.4 million),[9] with employment ranging from 5–50 people. Medium enterprises raked in annual revenue of LE10–100 million (USD1.4–14 million), with employment of 50–200 people. The World Economic Forum estimates that the overall contribution of SMEs to Egypt's economy is 38 percent of employment and 33 percent of GDP.[10]

Microenterprises

Microenterprises and very small enterprises, although sometimes lumped under the same category as SMEs (when referred to as MSMEs), tend to demonstrate slightly different characteristics. The importance of microenterprises was popularized by Muhammad Yunus and the Grameen Bank, as it exemplified the success of microfinance in Bangladesh.[11] Microenterprises tend to be extremely small operations, with fewer than five employees and revenue often at the subsistence level.

Outside of academia, the microenterprise phenomenon has grown in both policy and social support circles. Microenterprises have been noted to make up a significant portion of employment in lower-income economies. They often provide very important sources of income for lower-income populations, and generally are viewed as important alleviators of poverty.[12]

In Egypt, microenterprises may be defined as enterprises generating less than LE100,000 in revenue with employment ranging from one to four people.[13] Although those organizations are most vulnerable to external shocks and may not survive due to a lack of support, they have large potential to grow into sustainable small businesses if provided with the right support.[14] Almost 93 percent of firms in Egypt have fewer than four employees and make less than LE100,000 (USD14,000) in annual revenue.[15] Microenterprises that employ fewer than five employees contribute 72 percent of the employment of the non-agriculture private sector, which is around 20 percent of Egypt's overall employment.[16] Despite this big contribution to employment and number of firms, World Bank estimates indicate that microfinance outreach to those microenterprises is only 10 percent, leaving a huge untapped potential.[17]

High-growth entrepreneurship

According to the Organization for Economic Cooperation and Development (OECD), "all enterprises with average annualized growth greater than 20 percent per annum over a three year period should be considered as high-growth enterprises. Growth can be measured by the number of employees or by turnover."[18] In Egypt, available data for high-growth entrepreneurship is limited. The Global Entrepreneurship Monitor (GEM) studies of entrepreneurship, which measure a country's TEA in startup and new firms (up to 3.5 years old), can help estimate the size

Table 5.1: Growth and innovation potential for nascent and baby businesses

Country	TEA: Expected job growth ≥ 10 persons and ≥ 50 percent, in 5 years (% of TEA)	TEA: New product market combination (% of TEA)	TEA: Technology sector (medium-high and high-tech sectors) (% of TEA)
Factor-driven economies			
Egypt	0.71	14.64	1.04
Saudi Arabia	2.10	15.44	1.52
Pakistan	0.10	32.08	1.25
Vanuatu	0.33	34.88	3.76
Iran	0.77	14.41	3.07
Efficiency-driven economies			
Brazil	1.12	10.29	5.99
Turkey	1.50	26.37	2.60
South Africa	1.39	35.67	1.61
Innovation-driven economies			
United States	1.16	27.85	10.37
United Kingdom	0.64	23.46	11.58
France	0.57	33.80	5.09
Israel	1.03	24.39	3.13

Source: Data from Hattab, *Global Entrepreneurship Monitor: Egypt Entrepreneurship Report 2012.*

of high-growth entrepreneurship. From GEM data, we use indicators of innovation potential and growth in a firm's employment as proxies for opportunity entrepreneurship. As shown in table 5.1, in Egypt only 0.71 percent of total entrepreneurial activity is expected to create more than ten jobs and grow by 50 percent or more in five years.[19] This small percentage is comparable across different economies and is generally characteristic of high-growth/high-impact startups, which tend to be fewer in number but with significant impact; several studies suggest that a small number (3–10 percent) of fast-growing firms deliver substantial economic impact (50–80 percent of economic growth figures).[20] Overall, though, Egypt still has a lower rate in high-growth startups compared to markets such as Saudi Arabia, Turkey, South Africa, Brazil, and Israel.

Looking instead at measures of TEA related to pursuing product/ market innovation and/or launching a business in the medium and high-tech sectors—both characteristics of high-growth enterprises—these two proxies rank low relative to the global scale. Taking this insight to the firm level, table 5.1 suggests that Egypt performs worse when it comes to a new product/market combination (introducing unfamiliar product to selected market) compared with other factor-driven economies. This measure is placed at 14 percent of Egypt's TEA. Compared with other economies such as Pakistan (32 percent), Turkey (26 percent), and the United States (29 percent), Egypt ranks quite low. In addition, only 1 percent of Egypt's TEA is in the medium- to high-tech sector, which places Egypt toward the bottom of GEM-surveyed countries across multiple economic levels. On the macro level, the World Economic Forum's Global Competitiveness Report (2013–14) ranks Egypt 123 of 148 countries in terms of company spending on research and development,[21] a proxy that can be seen as representing levels of internal innovation.

Social entrepreneurship

A third category of entrepreneurs in Egypt and across the globe can be referred to as 'social' entrepreneurs. A relatively new concept, social entrepreneurship is associated with Ashoka, one of the world's leading institutions in promoting and supporting it. According to Ashoka, social entrepreneurs are individuals with innovative solutions to social problems.[22] The Skoll Foundation defines a social entrepreneur as someone who recognizes a social problem and uses entrepreneurial principles to organize, create, and manage a venture to make social change.[23]

The motivation for social entrepreneurship is thus often quite different from that of business entrepreneurship, although financial returns may factor into the equation. Social entrepreneurs' contribution to country economies is difficult to measure, since the field is still evolving and is in many cases unmapped. Aside from a few outliers, they do not tend to exhibit rapid growth; however, this could be because the field is new, untested, and does not operate within traditional markets. Facing a number of market inefficiencies in the issues they are trying to address, these organizations also tend to be less competitive. Their models may or may not be business oriented, and many demonstrate high levels of innovation.[24] While social entrepreneurs may not immediately affect a country's GDP, their positive effects on alleviating social challenges may in the longer term contribute to overall economic competitiveness and growth. This has been largely untested so far.

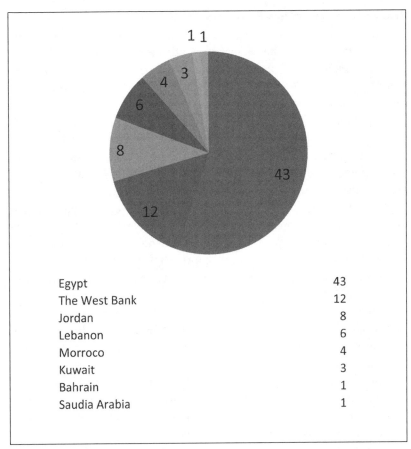

Egypt	43
The West Bank	12
Jordan	8
Lebanon	6
Morroco	4
Kuwait	3
Bahrain	1
Saudia Arabia	1

Figure 5.2 Internationally recognized Arab social entrepreneurs, by country

In Egypt, little data is available on the social entrepreneurial sector. Ehaab Abdou[25] identified forty-three internationally recognized Egyptian social entrepreneurs,[26] the highest share in the MENA region (55 percent). Since Egypt has the largest population in the Arab world, as well as some of the worst social problems, this may not necessarily indicate a higher level of social entrepreneurial activity, however. Based on the authors' experiences, we can reasonably estimate that social entrepreneurship, on balance, represents a small but growing share of Egyptian economic and entrepreneurial activity. Among the key NGOs promoting social entrepreneurship in Egypt are Nahdet El Mahrousa, Ashoka Arab World, Synergos, and, more recently, Misr El Kheir.

The Entrepreneurship Ecosystem in Egypt

An enabling ecosystem is necessary to raise the level of entrepreneurial activity. Studying the factors that enable the startup and growth of entrepreneurship is important for the development of policies and programs that can boost it, and by extension boost job creation, economic growth, and competitiveness. According to a World Economic Forum (WEF) study performed on the global level, the most important four factors in accelerating the growth of early-stage firms were top management, human resources, market opportunity, and access to funding and finance.[27] The most significant challenges were funding and finance, human resources, market opportunity, and government regulation.[28] It is worth noting that human resources are identified as both a factor for growth and a challenge. This matches the reality in Egypt, where entrepreneurs are often challenged to attract top talent, although the country produces substantial numbers of university graduates, especially engineers, who are often associated with technology-driven startups.

The German Development Institute (GDI) prepared a case in 2013 examining the characteristics of Egyptian SMEs that had upgraded to the status of 'gazelle' (high-growth) firms,[29] classifying the key factors studied under four blocks: entrepreneur characteristics, firm characteristics, business linkage to the value chain, and business environment. The results were generally consistent with WEF insights on the top accelerating and constraining factors on the growth of early-stage firms. For the entrepreneur, education and work experience were identified as the most important factors for upgrading, with 59 percent of interviewees highlighting them as key success factors. Indeed, 30 percent of 'gazelle' firm owners hold Master's degrees, compared to only 3 percent of non-upgraders. Forty-five percent of the interviewed SMEs considered work experience to be an important factor for upgrading. For a firm itself, the study showed that market research and innovation management, in addition to human resource development, were key determining factors for upgrading. Twenty-five percent of surveyed 'gazelles' showed that the majority of their workers had formal vocational training, while only 6 percent of non-upgraders had a majority of trained workers. For business environment, the factors that had the most impact as identified by interviewed SMEs were access to finance and the enforcement of the rule of law.

Entrepreneur havens

According to the WEF report *Accelerating Entrepreneurship*, there are currently about 150 initiatives that encourage entrepreneurship in the MENA region.[30] When examining the overall level of ecosystem support for entrepreneurship, the WEF classifies countries into three categories: "entrepreneur heavens," "late adopters," and "laggards." Entrepreneur heavens are defined as those that "have a large number of entrepreneurship support initiatives compared to the population."[31] This category includes countries such as Lebanon, Jordan, the UAE, Qatar, Oman, and Bahrain. Late adopters represent countries that have exerted efforts to "launch new initiatives to support entrepreneurs, but the number of initiatives are still low compared to population." [32] This category includes countries such as Saudi Arabia, Tunisia, and Morocco. Laggards are countries with very low rates of initiatives relative to their population, such as Egypt, Kuwait, Algeria, and Yemen.

Turning to Egypt, in the following section we examine these aspects of its entrepreneurship ecosystem:

- Access to finance
- Education and training
- Existing entrepreneurship support systems
- Research and development (R&D) transfer
- The regulatory framework related to ease of doing business

Access to finance

High-growth entrepreneurship is often contingent on financial support.[33] According to the 2012 GEM report, expert opinions suggest that current financial support is not sufficiently meeting the demand of the entrepreneurial landscape in Egypt, which experts rank 38 out of 69 economies, with an average score of 2.38 on a scale of 1–5 (1 and 2 are negative for entrepreneurship, 3 is average, while 4 and 5 are positive).[34] This result consolidates the experts' overall opinion on the various available funding mechanisms found in Egypt, including government funding, venture capital funding, debt funding, and individual equity funding such as angel investors and IPOs (table 5.2).

Recently, several initiatives and organizations have been or are being launched to cover a portion of this gap, including venture capital funds, angel investor networks, and specialized banking programs. The sections below provide an overview of the funding supply side and its characteristics in Egypt.

Table 5.2. Experts' opinion on funding-support environment

Assessment components	Score (1–5)
Sufficient funding available through initial public offerings for new and growing firms	1.9
Sufficient venture capitalist funding available for new and growing firms	2.6
Sufficient funding available from private individuals (other than founders) for new and growing firms	2.6
Sufficient government subsidies available for new and growing firms	2.1
Sufficient debt funding available for new and growing firms	2.6
Sufficient equity funding available for new and growing firms	2.5

Source: Data from Hattab, *Global Entrepreneurship Monitor: Egypt Entrepreneurship Report 2012*

Family and personal funds

Family and personal funds tend to be the primary source of financial support for Egyptian entrepreneurs.[35] Personal savings in particular are cited as the most significant financial investment by Egyptian entrepreneurs, used by 72 percent of nascent entrepreneurs and 68 percent of new firms in starting and growing their businesses. This is followed by family funds (19 percent) for both nascent and new firms.[36] The German Development Institute (GDI) notes that entrepreneurs' personal funds played a major role in firms that were able to upgrade to high-growth firms.[37] Accordingly, we can conclude that the lack of funding sources other than personal and family savings plays a limiting role in growing entrepreneurship and upgrading SMEs in Egypt.

Private investments

Beyond personal and family funds, other sources of funding in Egypt include venture capital, angel investors, and crowdfunding. For new firms, this type of capital makes up 3 percent of total investment, while it does not represent anything for nascent entrepreneurs (0 percent).[38]

The concept of venture capital (VC) is still new in the MENA region. According to a 2013 MENA Private Equity Association report, the number of completed VC transactions in the region was 119 in 2010–2012.[39] Countries such as Tunisia, Morocco, and Lebanon lead in terms of total VC transactions in 2010–2012 (27, 26, and 25), and while growth in VC investments in Egypt (from only 4 in 2007–2009, to 15 in 2010–2012) is consistent with regional MENA trends (from 56 transactions in 2007–2009, to 119 in 2010–2012), the total number of transactions in Egypt is still minimal. In the MENA region, from a funding and transaction value standpoint, from 2011 to 2012 the number of VC firms that had raised funds increased from 9 to 11, yet the total amount of investment decreased from usd612 million to usd308 million. This decrease is attributed to the effect of the global financial crisis in addition to the political instability in key markets such as Egypt.[40]

Among Egypt's oldest and largest VC firms is Ideaveloper, the VC arm of EFG-Hermes. Other VCs include Sawari Ventures, which focuses on the technology, media, and telecommunications sectors; Orascom Telecom Venture (the venture capital arm of Orascom Telecom); and Nile Capital/IT Venture. For high-growth and pre-IPO stages, there are also several private equity firms in Egypt, such as Citadel Capital and Naeem Capital. Private equity firms look for highly profitable business potential and a quick exit through IPOs or different means.

Angel investors, or individuals who provide financial backing from personal funds, offer an additional form of informal financing for startups. In Egypt, the angel investor profile is diversified, including businessmen and women with generated wealth, successful entrepreneurs, and senior managers from multinationals and large corporations. For instance, a newly initiated network of angel investors, named Cairo Angels, collaborates in assessing potential business opportunities and makes individual equity investments in the range of le250,000 to le1 million (usd35,000–140,000), aiming to increase the overall level of angel investments in Egypt.[41]

A third form of funding has recently been introduced in Egypt: crowdfunding. Crowdfunding is the practice of funding a project or venture by raising small, individual sums of money from large groups of people, mostly through the Internet. For instance, Yomken[42] is a newly-established organization in Egypt that provides this type of funding to micro and small entrepreneurs, focusing on the low-tech industry; Shekra[43] is a crowdfunding organization working with entrepreneurs at different stages; and Madad[44] is a crowdfunding portal focusing on development projects. Crowdfunding is still in the infancy stages in Egypt and crowdfunding platforms are facing significant legal and institutional challenges.

Bank financing

In Egypt, 47 percent of SMEs deal with banks, of which 22 percent have bank facilities.[45] For startup entrepreneurs, however, the story is different. In terms of obtaining bank funding for entrepreneurial activity, the 2012 Egypt GEM report found that only 6 percent of nascent firms and 7 percent of new firms raise funds from banking and other financial institutions in Egypt.[46] The smaller the firm's sales turnover, the less likely it is to have banking facilities: only 12 percent of companies with less than LE0.5 million in sales turnover have facilities from banks, while 56 percent of companies with a sales turnover of LE20–50 million do.[47] Also, firms with lower capital tend to have less access to bank facilities: only 18 percent of firms with capital of less than LE250,000 have access to bank facilities, while 58 percent of firms with capital of LE30 million and above do, suggesting that the smaller the firm, the less likely it is to have access to banking facilities.[48]

There are several major gaps that prevent entrepreneurs from obtaining commercial bank facilities to start or upgrade their businesses. Conservative and antiquated bankruptcy laws elevate the risk for entrepreneurs and create a culture of mistrust on the part of banks toward entrepreneurs, while entrepreneurs tend to lack adequate knowledge of the banking sector and their organizations lack the internal capability to complete bank requirements for raising debt financing. In fact, many entrepreneurs are fearful of obtaining bank facilities due to the threat of prosecution and even jail if they cannot pay outstanding debt.[49]

Some NGOs, such as the Egyptian Banking Institute (EBI), play a role in bridging the gap between financial institutions and entrepreneurs. The EBI is funded by the Central Bank of Egypt (CBE), and it encourages Egyptian financial competitiveness by providing more than 16,000 professionals annually with a range of educational and training programs on finance and banking.[50] One example is a program in which eight HSBC customers were provided with financial training, as a result of which four obtained HSBC financing.[51] This suggests a way in which financial education through NGOs could bridge part of the gap between commercial banks and entrepreneurs.

Education and training

Entrepreneurship training and its integration into educational systems, according to the GEM, has been rated as very low in Egypt, with a score 1.28 on scale of 1–5.[52] In comparison with GEM-surveyed economies,

Egypt ranked 69, at the bottom of the list. GEM expert surveys suggest the country has low levels of educational services in terms of entrepreneurship training. From the primary and secondary levels, instruction in entrepreneurship, market economics, creativity, self-sufficiency, and personal initiative is cited as low. At the tertiary, vocational, and professional education levels, education on business startups and management is also ranked poorly. Together, these indicators reveal a significant gap in educational preparedness for entrepreneurship in Egypt (table 5.3).

In the authors' experience, some private universities, such as the American University in Cairo (AUC), the German University in Cairo (GUC), the British University in Egypt (BUE), and the Arab Academy for Science and Technology (AAST), provide students with opportunities for entrepreneurial education, either by integrating it into the core curriculum or by promoting and fostering extracurricular activities. For instance, AUC has established the AUC Venture Lab (an accelerator/incubator) and launched the Women Entrepreneurship and Leadership (WEL) Program, neither of which is exclusive to AUC students. Most of the training currently provided is focused on soft skills and leadership development.[53] Although this plays a major role in developing students' capabilities and awareness, more focused training is needed to promote entrepreneurship and the concept of creating new ventures. Student organizations that work with the student community to develop entrepreneurship include

Table 5.3. Experts' opinion on entrepreneurship education and training

Training Assessment Components	Score (1–5)
Vocational, professional, and continuing education systems provide good and adequate preparation for starting up and growing new firms	1.82
The level of business and management education provides good and adequate preparation for starting up and growing new firms	1.88
Colleges and universities provide good and adequate preparation for starting up and growing new firms	1.92
Teaching in primary and secondary education provides adequate attention to entrepreneurship and new firm creation	1.18
Teaching in primary and secondary education provides adequate instruction in market economics principles	1.33
Teaching in primary and secondary education encourages creativity, self-sufficiency, and personal initiative	1.32

Source: Data from Hattab, *Global Entrepreneurship Monitor: Egypt Entrepreneurship Report 2012*

Injaz Egypt and Enactus.[54] Both work within the university system to conduct training and business competitions, enabling students to practice entrepreneurial skills in the form of extracurricular activities. On balance, programs like these aim to upgrade to an educational culture that can enable and promote entrepreneurship.

Support systems: incubation and mentorship

A number of organizations external to formal education have begun to pop up in support of entrepreneurship education, training, and preparation in Egypt. These include incubators/accelerators, professional/advisory services, mentorship programs, non-governmental organizations, and governmental programs that support entrepreneurs.

As defined by the European Commission,[55] a business incubator is an organization that accelerates and systematizes the process of creating successful enterprises by providing them with a comprehensive and integrated range of support aimed at addressing multiple entrepreneurial challenges, including access to financing, training programs, professional services, mentorship programs, clustering and networking opportunities, and working space. The GEM's 2012 Egypt report ranked incubators' lack of nonfinancial and business support (either through incubators or dedicated programs) as the third most important entrepreneurship-constraining factor, coming after lack of access to financial solutions and insufficient formal and informal education systems promoting entrepreneurship and creativity.

Figure 5.3 shows the results of a survey conducted by the authors of a sample of twenty-four incubated startups and two of the main incubators in Egypt to assess the impact of incubators and startups' overall satisfaction with incubator services. The survey was based on a scale of 1–5 (below 3 is a negative response, 3 is average, and above 3 is positive).

Results indicated that startups' satisfaction with surveyed incubator-provided services was average, with the exception of access to market research services, which was rated below average. Support in recruitment and networking with similar businesses were rated as borderline satisfactory. Access to office space was the highest-rated service. This high satisfaction is explained primarily by the ease of providing tangible physical services versus more sophisticated services such as coaching and networking.

Conversely, in the 2012 Egypt GEM report, experts' opinions rated professional services with a score of 2.6 out of 5 (66 of 69 countries surveyed), indicating these services were below average on meeting

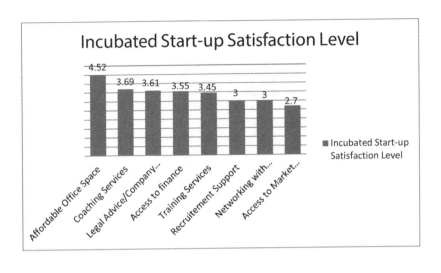

Affordable Office Space	4.52
Coaching Services	3.69
Legal Advice/Company Registeration	3.61
Access to finance	3.55
Training Services	3.45
Recruitement Support	3
Networking with similar business	3
Access to Market Research	2.7

Figure 5.3 Incubator services satisfaction perception

entrepreneurs' demand for support.[56] While these numbers slightly contrast with the results from our survey, we can surmise that while current
incubator services seem to rank positively in terms of the support they
provide, the level, scope, and, most important, reach of these services may
not be adequate for the overall entrepreneurship ecosystem across the
country. Government support programs were ranked even lower by the
GEM expert surveys, with a score of 1.86 out of 5 (64 of 69 countries surveyed). While the General Authority for Investment (GAFI) is offering
a "one-stop-shop" (to simplify startup procedures) and a Social Fund for
Development, experts believe these efforts are still insufficient to meet
the size of demand and need in Egypt. Table 5.4 shows a sample of the
known and established incubators/accelerators in Egypt, as well as other
supporting organizations.

Table 5.4. A sample of the known and established incubators/accelerators in Egypt, and other supporting organizations

Incubator	Description
AUC Venture Lab	The AUC Venture Lab (AUC V-Lab) is AUC's startup incubator based at AUC New Cairo. AUC V-Lab will enable startups to capitalize on AUC's world-class facilities and knowledge base, connecting innovative startups with AUC's alumni network and fostering a thriving ecosystem of innovation, education, and business. http://www.aucegypt.edu/Business/eip/Pages/Venture%20Lab.aspx
Flat6Labs	Flat6Labs fosters and invests in Egyptian technology startups for a three-month period, providing them with access to facilities, expertise, mentorship, and the support needed to launch into orbit and be ready to face the challenges of the local and global market. www.flat6labs.com
Innoventures	Innoventures' mission is to turn the creative ideas of our entrepreneurs into successful businesses that change the world. www.innoventures.me
Tamkeen Capital	Tamkeen Capital is a venture ally firm aspiring toward a sustainable business of innovation, founded in 2011, just after the Egyptian revolution.
TIEC	The Technology Innovation and Entrepreneurship Center (TIEC) is a government-led incubator aiming to drive innovation and entrepreneurship in information and communications technology for the benefit of the national economy. The center was launched at Smart Village in September 2010. www.tiec.gov.eg
Endeavor	Headquartered in New York, Endeavor is an international non-profit organization that seeks to transform emerging markets by supporting entrepreneurship. In 2008, Endeavor opened affiliates in Egypt, Jordan, and India. www.endeavoreg.org

Source: Data from authors.

Research and development, and technology transfer

As highlighted in previous sections, innovation and research and development (R&D) are central to innovation-driven economies as enablers for high-growth entrepreneurship. Egypt GEM experts ranked R&D transfer in Egypt with score of 1.83 on a scale of 1–5 (ranked 68 out of 69 GEM-surveyed countries). These results consolidate the experts' overall assessment and opinion of several R&D aspects in Egypt. Most of the assessed aspects were rated below 2, as shown in table 5.5.

Table 5.5. Experts' opinion on R&D transfer in Egypt

R&D Enablers Assessment Components	Score (1–5)
There is good support available for engineers and scientists to have their ideas commercialized through new and growing firms.	1.67
The science and technology base efficiently supports the creation of world-class, new technology-based ventures in at least one area.	1.86
There are adequate government subsidies for new and growing firms to acquire new technology.	1.83
New and growing firms can afford the latest technology.	2.28
New and growing firms have just as much access to new research and technology as large, established firms.	1.91
New technology, science, and other knowledge are efficiently transferred from universities and public research centers to new and growing firms.	1.50

Source: Data from Hattab, *Global Entrepreneurship Monitor: Egypt Entrepreneurship Report 2012*

Legal and regulatory framework

The presence of a legal regulatory framework friendly to entrepreneurship is crucial to upgrading entrepreneurial firms. Its laws protect property rights; facilitate flexible contractual relations; facilitate international transactions; enable different types of investments such as venture capital, angel investments, and crowdfunding; and provide bankruptcy protection for both entrepreneurs and investors. Additionally, an effective regulatory framework provides for a rapid and just process for litigation, and strict enforcement of court rulings. Egypt has a cumbersome regulatory regime that has not been modernized, which presents many obstacles for doing business in general and for entrepreneurs in particular. For example, incorporation and contractual laws make it hard to enact certain minority protections for investors, or to create stock option plans. Old bankruptcy laws raise the risk for entrepreneurs and encourage them to become more risk averse in their growth and investments. Crowdfunding mechanisms are also unable to operate under current laws (which is a common problem in most countries). Social enterprises that aim for a social mission while still operating as a for-profit entity face obstacles between the NGO law and corporate laws. The regulatory framework needs to be reviewed and modernized to allow for modern arrangements between entrepreneurs, investors, and other stakeholders. Additionally, the litigation process and enforcement must be improved.

The impact of these limitations in the regulatory regime is visible in the "doing business" environment. According to the World Bank's Doing Business indicators, Egypt ranks 128 of 189 economies worldwide (the average for the Middle East and North Africa region is 107).[57] Table 5.6 shows the regional breakdown of the ranking on different aspects of doing business. As shown, Egypt ranks above the regional average for ease of starting business procedures, access to credit, and trading across borders, though most of these scores are still lower than those of the UAE, the leading player in the Middle East region. Aspects ranking below the regional average that need special focus include contract enforcement in case of disputes, tax paying, investor protection, insolvency resolution, and access to infrastructure such as electricity and property registration.

Table 5.6: Ease of doing business in Egypt and the region

Indicator	Country ranking							
	Egypt	Jordan	Lebanon	Saudi Arabia	Syria	Turkey	UAE	Regional average
Starting a business	50	117	120	84	135	95	37	112
Trading across borders	83	57	97	69	147	86	4	89
Getting credit	86	170	109	55	180	86	86	113
Getting electricity	105	41	51	15	82	49	4	77
Registering property	105	104	112	14	82	50	4	93
Resolving insolvency	146	113	93	106	120	130	101	105
Protecting investors	147	170	98	22	115	34	98	113
Paying taxes	148	35	39	3	120	71	1	64
Dealing with construction permits	149	111	179	17	189	148	5	108
Enforcing contracts	156	133	126	127	179	38	100	118

Source: Data from World Bank and the International Finance Corporation, *Doing Business: Regional Profile Middle East and North Africa*

Conclusion

The entrepreneurship landscape across the globe can be broadly characterized by three types of entrepreneurs: necessity, opportunity, and social. Egypt is heavily dominated by necessity entrepreneurs, particularly at the micro and small enterprise levels. These entrepreneurs face a host of challenges: financial, regulatory, human capital, and cultural. Together, these challenges limit the ability of entrepreneurs to start up or achieve growth on a scale that can significantly benefit large swaths of the population. Based on our review of the country's entrepreneurship ecosystem, we highlight three areas that are needed to promote entrepreneurship in Egypt.

Encouraging opportunity-based entrepreneurship and upgrading necessity-based entrepreneurship

On strategic and macro levels, policymakers should design programs that enable more opportunity-based entrepreneurship by encouraging innovation and designing special programs that encourage these entrepreneurs to pursue the entrepreneurship track. At the same time, policymakers must also address necessity-based entrepreneurship, which is dominant in Egypt, by providing the needed programs and environment that enable these entrepreneurs to move up to the next level of growing their firms.

Focus on key ecosystem enablers

Access to finance, regulatory frameworks, and human capital have been highlighted as the most important and challenging aspects that can either hinder or accelerate the growth of startups. Special focus from policymakers is required to tackle them. For instance, educational systems need to be revamped by integrating the concept of entrepreneurship in curricula, as well as in formal and informal academic activities. As for access to finance, policymakers should promote existing programs and enable new ones to help bridge the gap between startups and banks, in addition to fostering nontraditional funding sources such as venture capital and angel investors.

Focus on enabling innovation and research and development

Enabling a climate for innovation, R&D development, and support initiatives is crucial to drive Egypt's economy from being factor-driven to becoming innovation-driven, and to support opportunity-driven startups in upgrading to high-growth ventures. This will not come about without an integration of different ecosystem enablers that are centered on innovation and R&D.

Notes

1   Niels Bosma, Sander Wennekers, and José Ernesto Amorós, *Global Entrepreneurship Monitor: 2011 Extended Report: Entrepreneurs and Entrepreneurial Employees across the Globe* (Babson Park, Santiago, Kuala Lumpur, London: Global Entrepreneurship Research Association (GERA), 2011), 19.

2   Siri Roland Xavier et al., *Global Entrepreneurship Monitor: 2012 Global Report* (Babson Park, Santiago, Kuala Lumpur, London: Global Entrepreneurship Research Association (GERA), 2012).

3   World Economic Forum, *The Global Competitiveness Report 2013–2014* (World Economic Forum, 2013).

4   Siri Roland Xavier et al., *Global Entrepreneurship Monitor: 2012 Global Report.*

5   Tahseen Consulting, "Promoting Entrepreneurship in the Arab World: The Need for Tailored National Approach," Tahseen Consulting Blog, 2013, http://tahseen.ae/blog/?p=705&goback=.gde_4578702_member_275579935#.

6   Siri Roland Xavier et al., *Global Entrepreneurship Monitor: 2012 Global Report.*

7   Tom Gibson and H.J. Van der Vaart, *Defining SMEs: A Less Imperfect Way of Defining Small and Medium Enterprises in Developing Countries* (Brookings Global Economy and Development, 2008).

8   Khrystyna Kushnir, Melina Laura Mirmulstein, and Rita Ramalho, *Micro, Small, and Medium Enterprises around the World: How Many Are There, and What Affects the Count* (Washington DC: World Bank/IFC MSME Country Indicators Analysis Note, 2010).

9   Robert Poldermans, *Expanding Egypt's Banking Frontiers: The Future of SME Banking in Egypt* (Central Bank of Egypt and Egyptian Banking Institute, 2011).

10 Booz & Company, *Accelerating Entrepreneurship in the Arab World* (World Economic Forum in collaboration with Booz & Company, 2011).

11 Dharam P. Ghai, "An Evaluation of the Impact of the Grameen Bank Project," Grameen Bank, 1985.

12 Susan Johnson and Ben Rogaly, *Microfinance and Poverty Reduction* (Oxfam, 1997); Shahidur Khandker, *Micro-finance and Poverty: Evidence Using Panel Data from Bangladesh* (World Bank, Development Research Group, 2003).

13 Poldermans, *Expanding Egypt's Banking Frontiers.*

14 Booz & Company, *Accelerating Entrepreneurship in the Arab World.*

15 Poldermans, *Expanding Egypt's Banking Frontiers.*

16 Hafez Ghanem. *The Role of Micro and Small Enterprises in Egypt's Economic Transition* (Washington, DC: The Brookings Institution, 2013).

17 Sahar Nasr, *Access to Finance and Economic Growth in Egypt* (The World Bank Group, 2007).

18 OECD, *High Growth Enterprises*, Eurostat-OECD Manual on Business Demography Statistics, 2014, http://www.oecd.org/industry/business-stats/39974588.pdf.

19 Hala Hattab, *Global Entrepreneurship Monitor: Egypt Entrepreneurship Report 2010* (GEM Consortium, 2010).

20 Mike Ducker, "Egypt Entrepreneurship Final Report: Where Are All the Egyptian Entrepreneurs?" (J.E. Austin Associates Inc., in collaboration with Deloitte, 2009.) http://www.academia.edu/5193052/egypt_entrepreneurship_final_report_where_are_all_the_egyptian_entrepreneur

21 WEF, *The Global Competitiveness Report.*

22   "What is a Social Entrepreneur?" https://www.ashoka.org/social_entrepreneur.

23   Patrick O'Heffernan, *Defining Social Entrepreneurship*, Skoll World Forum, 2007, http://skollworldforum.org/2007/07/10/defining-social-entrepreneurship/.

24   Siri Terjesen et al., "Report on Social Entrepreneurship" (Global Entrepreneurship Monitor, 2009).

25   Ehaab Abdou is a social entrepreneur and researcher. He cofounded a number of NGOs including Nahdet El Mahrousa, Fathet Kheir, and Ana Masri. In 2004, he was selected as an Ashoka fellow. He conducted and published research at the World Bank Institute and the Brookings Foundation.

26   Ehaab Abdou et al., "Social Entrepreneurship in the Middle East: Toward Sustainable Development for the Next Generation" (Middle East Youth Initiative, 2012).

27   World Economic Forum, *Entrepreneurial Ecosystems around the Globe and Company Growth Dynamics* (World Economic Forum, 2013).

28   World Economic Forum, *Entrepreneurial Ecosystems around the Globe and Company Growth Dynamics* (World Economic Forum, 2013).

29   German Development Institute, "The Case of Egypt: Which Factors Determine the Upgrading of Small and Medium-Sized Enterprises (SMEs)?" (German Development Institute, 2013).

30   Booz & Company, *Accelerating Entrepreneurship in the Arab World.*

31   Booz & Company, *Accelerating Entrepreneurship in the Arab World.*

32   Booz & Company, *Accelerating Entrepreneurship in the Arab World.*

33   World Economic Forum, *Entrepreneurial Ecosystems around the Globe and Company Growth Dynamics.*

34   Siri Roland Xavier et al., *Global Entrepreneurship Monitor: 2012 Global Report.*

35   Hala Hattab, *Global Entrepreneurship Monitor: Egypt Entrepreneurship Report 2012.*

36   Hattab, *Global Entrepreneurship Monitor: Egypt Entrepreneurship Report 2012.*

37   German Development Institute, "The Case of Egypt."

38   Hattab, *Global Entrepreneurship Monitor: Egypt Entrepreneurship Report 2012.*

39   MENA Private Equity Association, *Third Venture Capital in Middle East and North Africa*, 2013, http://static.wamda.com/web/uploads/resources/VC_report_2013_Final_1.pdf

40   MENA Private Equity Association, *Third Venture Capital in Middle East and North Africa.*

41   Cairo Angels, "Cairo Angels: A Forum for Investors to Find High Potential Start Ups." http://www.cairoangels.com/.

42   See www.yomken.com for more information.

43   See www.shekra.com for more information.

44   See www.madad.com.eg for more information.

45   Hand El Said, Mahmoud Al Said, and Chahir Zaki, "Small and Medium Enterprises Landscape in Egypt: New Facts from a New Dataset," *Egyptian Banking Institute*, 2011, http://www.sme-egypt.com/.

46   Hattab, *Global Entrepreneurship Monitor: Egypt Entrepreneurship Report 2012.*

47   El Said et al., "Small and Medium Enterprises Landscape."

48   El Said et al., "Small and Medium Enterprises Landscape."

49   Ducker, "Egypt Entrepreneurship Final Report."

50   For more information, see http://www.ebi.gov.eg/.

51   Ducker, "Egypt Entrepreneurship Final Report."

52 Hattab, *Global Entrepreneurship Monitor: Egypt Entrepreneurship Report 2010.*
53 Ducker, "Egypt Entrepreneurship Final Report."
54 For more information, see www.injaz-egypt.org and www.enactus.org
55 OECD, *High Growth Enterprises.*
56 Hattab, *Global Entrepreneurship Monitor: Egypt Entrepreneurship Report 2012.*
57 World Bank and the International Finance Corporation, *Doing Business: Regional Profile Middle East and North Africa* (Washington, DC: International Bank for Reconstruction and Development/World Bank, 2014).

Bibliography

Abdou, Ehaab, Amina Fahmy, Dinana Greenwald, and Jane Nelson. "Social Entrepreneurship in the Middle East: Toward Sustainable Development for the Next Generation." Middle East Youth Initiative, 2012.

Ashoka. "What is a Social Entrepreneur?" https://www.ashoka.org/social_entrepreneur

Booz & Company. *Accelerating Entrepreneurship in the Arab World.* World Economic Forum in collaboration with Booz & Company, 2011.

Bosma, Niels, Sander Wennekers, and José Ernesto Amorós. *Global Entrepreneurship Monitor: 2011 Extended Report: Entrepreneurs and Entrepreneurial Employees across the Globe.* Babson Park, Santiago, London: Global Entrepreneurship Research Association (GERA), 2011.

Cairo Angels. "Cairo Angels: A Forum for Investors to Find High Potential Start Ups." http://www.cairoangels.com/

Ducker, Mike. "Egypt Entrepreneurship Final Report: Where Are All the Egyptian Entrepreneurs?" J.E. Austin Associates Inc. in collaboration with Deloitte, 2009. http://www.academia.edu/5193052/egypt_entrepreneurship_final_report_where_are_all_the_egyptian_entrepreneurs

German Development Institute. "The Case of Egypt: Which Factors Determine the Upgrading of Small and Medium-Sized Enterprises (SMEs)?" German Development Institute, 2013.

Ghai, Dharam P. "An Evaluation of the Impact of the Grameen Bank Project." Grameen Bank, 1985.

Ghanem, Hafez. *The Role of Micro and Small Enterprises in Egypt's Economic Transition.* Washington D.C.: The Brookings Institution, 2013.

Gibson, Tom, and H.J. Van der Vaart. *Defining SMEs: A Less Imperfect Way of Defining Small and Medium Enterprises in Developing Countries.* Brookings Global Economy and Development, 2008.

Hattab, Hala. *Global Entrepreneurship Monitor: Egypt Entrepreneurship Report 2010.* GEM Consortium, 2010.

———. *Global Entrepreneurship Monitor: Egypt Entrepreneurship Report 2012.* GEM Consortium, 2012.

Kelley, Donna, Niels Bosma, and José Ernesto Amorós. *Global Entrepreneurship Monitor: 2010 Global Report.* Babson Park, Santiago, London: Global Entrepreneurship Research Association (GERA), 2010.

Khandker, Shahidur. *Micro-finance and Poverty: Evidence Using Panel Data from Bangladesh.* Washington DC: World Bank, Development Research Group, 2003.

Kushnir, Khrystyna, Melina Laura Mirmulstein, and Rita Ramalho. *Micro, Small, and Medium Enterprises around the World: How Many Are There, and What Affects the*

Count. Washington, D.C.: World Bank/IFC MSME Country Indicators Analysis Note, 2010.

Johnson, Susan, and Ben Rogaly. *Microfinance and Poverty Reduction.* Oxfam, 1997.

MENA Private Equity Association. *Third Venture Capital in Middle East and North Africa.* 2013. http://static.wamda.com/web/uploads/resources/VC_report_2013_Final_1.pdf

Nasr, Sahar. *Access to Finance and Economic Growth in Egypt.* The World Bank Group, 2007.

OECD. *High Growth Enterprises.* Eurostat-OECD Manual on Business Demography Statistics. 2014. http://www.oecd.org/industry/business-stats/39974588.pdf

O'Heffernan, Patrick. *Defining Social Entrepreneurship.* Skoll World Forum, 2007. http://skollworldforum.org/2007/07/10/defining-social-entrepreneurship/

Poldermans, Robert. *Expanding Egypt's Banking Frontiers: The Future of SME Banking in Egypt.* Central Bank of Egypt and Egyptian Banking Institute, 2011.

El Said, Hand, Mahmoud Al Said, and Chahir Zaki. "Small and Medium Enterprises Landscape in Egypt: New Facts from a New Dataset." *Egyptian Banking Institute,* 2011. http://www.sme-egypt.com/

Tahseen Consulting. "Promoting Entrepreneurship in the Arab World: The Need for Tailored National Approach." Tahseen Consulting Blog, 2013. http://tahseen.ae/blog/?p=705&goback=.gde_4578702_member_275579935#

Terjesen, Siri, Jan Lepoutre, Rachida Justo, and Niels Bosma. "Report on Social Entrepreneurship." Global Entrepreneurship Monitor, 2009.

World Economic Forum. *Entrepreneurial Ecosystems around the Globe and Company Growth Dynamics.* World Economic Forum, 2013.

———. *The Global Competitiveness Report 2013–2014.* World Economic Forum, 2013.

World Bank and the International Finance Corporation. *Doing Business: Smarter Regulations for Small and Medium-size Enterprises.* Washington, DC: International Bank for Reconstruction and Development/World Bank, 2013.

Xavier, Siri Roland, Donna Kelley, Jacqui Kew, Mike Herrington, and Arne Vorderwülbecke. *Global Entrepreneurship Monitor: 2012 Global Report.* Babson Park, Santiago, Kuala Lumpur, London: Global Entrepreneurship Research Association (GERA), 2012.

6. Entrepreneurs in the 'Missing Middle': Know Your Funding Options

Adel Boseli

Many entrepreneurs look at the word 'funding' and pause. Some see it as an obstacle; others see it as a black box. The reality is, once you know your funding options and learn how to best to utilize them, your business path becomes clearer. The purpose of this chapter is to act as a guide for Egyptian entrepreneurs, to help them identify their funding options and choose the most suitable ones for their needs. All examples in this chapter apply to the Egyptian market in 2014.

The Middle East and North Africa (MENA) region generally, and Egypt specifically, is still evolving in the area of entrepreneurship. Nevertheless, young entrepreneurs worldwide share several common challenges and characteristics. This chapter will focus on such commonalities related to funding. When it comes to funding, one of the main characteristics entrepreneurs share is falling into the 'missing middle.'

The 'Missing Middle'

Silatech and Gallup Inc. report that most entrepreneurial Arab youth identified availability and access to financing as the main barrier of entry for their ventures.[1] This is referred to as the 'missing middle,' a funding gap in the financing ecosystem. The 'missing middle' is an obstacle faced by both entrepreneurs and small to medium enterprises (SMEs).

The 'missing middle' refers to the lack of investable capital targeted at funding SMEs. This typically involves companies or startups that have

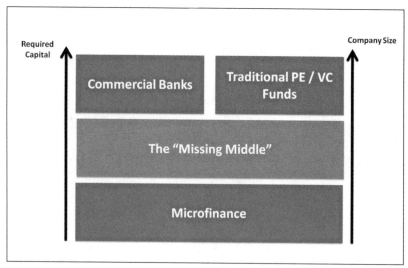

Figure 6.1 The 'missing middle'

specific financing needs that are too big for microfinance, too small for private equity funds, and too risky for banks[2] (see figure 6.1). On the one hand, microfinance, usually provided by NGOs and personal donations, is mainly used for poverty alleviation and necessity-driven entrepreneurs. On the other hand, big companies with significant financing needs resort to banks and private equity funds. Most entrepreneurs are positioned somewhere in between, however, and thus fall in an underserved area, coined the 'missing middle.'

Although at first glance these facts look alarming, over the last couple of years there has been positive change in this regard. There is an increased awareness of the potential of such enterprises and also their role in employment and innovation. As a result, more funding initiatives are channeled their way.

Investment Readiness

From the entrepreneur's side, before seeking funding or getting caught up in all the details of funding sources and choices, it is important to pose this question: Am I ready for investment? Determining whether a company is ready for investment is not always an easy task. A company might have a product with high customer demand or purchase orders, yet it may lack the funding to manufacture the product. In this case, seeking funding sounds like the right choice. On the other hand, another startup company might want to recruit core team members first, and thus decide to raise funds before there is demand for their product. Since the company is yet to have a high value in terms of product, the value of funding required would probably translate into a large part of the company's equity because from an investment perspective, it is too risky. Offering equity to high-caliber candidates to join, or waiting for the company to mature and thus be able to afford it, could be better options.

Although it can sometimes seem clear from the perspective of an outsider whether a company should seek investment, the decision is easier to make from within the company. Besides cases where funding is pursued out of necessity, there is always a struggle between seeking funding to grow the company and losing some equity versus bootstrapping[3] to be able to raise more funding for less equity later. There is no magic formula, but entrepreneurs must be aware of the consequences of the decision they make in this regard and must evaluate if and when the enterprise is ready for funding. The Investment Readiness Checklist provided here is compiled from a number of sources, including RBC Royal Bank,[4] New York Angels,[5] and the author's experience in the field:

Table 6.1. Investment readiness checklist

Area	Things to check
General	Can you write one sentence that shows your core value proposition?
	Are you ready to sell shares?
	Does the team contain experts in the industry?
	Does the team have management experience?
	Is the team diverse in experience?
	Does the company or team have references from previous work?
	Does the company or team have previous success stories to support their case?
	Do you have a realistic business plan?
	Are you willing to include investors in the management of the company?

	Do you have a clear growth strategy and product development plan?
	Do you have a staffing plan and know the required calibers and their cost?
	Do you have an advisory board?
	Do you know your risks and have a risk avoidance and mitigation plan?
	Is your intellectual property protected where needed?
	Have you prepared a simple and organized investment deck that summarizes the team, market, product, potential, and use of capital?
	Have you rehearsed and perfected your investors pitch?
	Are you ready for due diligence from investors?
Financial	Do you have an achievable financial forecast?
	Are you offering potential for high returns on investment (ROI)?
	Did you do a valuation for your business on solid grounds and do know how much equity you are offering investors?
	Do you have a tractable accounting record?
	Can you break down how the funding will be used?
	Have you decided how to manage returns (for example, dividends distribution and reinvesting in the company for growth)?
	Do you know the exit strategy you will provide investors?
	Did you plan your cash flow?
Market	Did you conduct any market research?
	Do you understand the target market and clearly know your target client?
	Is the target market big enough or does it have high potential?
	Do you know your competition well?
	Is there a barrier to entry to the market?

	Do you have the advantage of being first in market?
	Do you have a clear marketing and sales strategy?
	Do you have early sales or does the market need validation?
	Is there a global need for what you are offering?
Legal	Do you have a legal structure?
	Do you have a clear understanding of the governing laws and regulations affecting your business?
	Do you have a current shareholders agreement?
	Are employee contracts in place?

The above questions are merely meant to serve as guidelines and eye-openers. Moreover, the set of answers that indicate whether a company is ready for investment depends on a number of factors: the type of business, future vision, and most important, which stage the company or startup is in. For example, if the company is still in an early stage and needs investment to develop a prototype or hire core team members, the risk the investor will bear is high and there are fewer grounds to help fairly valuate the company. Accordingly, it is more likely to get a poor valuation and be forced to give up a lot of equity for investment. If an entrepreneur feels that he or she is unable to answer most of the above checklist, this indicates that further work is required. This will allow for a better valuation and a better chance of delivering an attractive pitch to investors.

Startup Stages

For the scope of this chapter, I classify the lifetime of a startup into five stages: pre-seed, seed/startup, early-stage, growth, and maturity. The pre-seed stage is when entrepreneurs are still exploring their business idea. They do not know whether their business idea is viable, or whether there is demand for their product, and they might not even know whether they can succeed. The vision of the company/startup might not be clear, and all team members might not be on board. At this stage, it is best to be funded by personal financing, whether from "friends, families, and fools" (FFF), accelerators, incubators, or business plan competitions. The only exceptions are research and development (R&D) projects that may pursue university grants or R&D funds. R&D projects are outside the scope of this chapter.

As the idea matures, and it looks more like the entrepreneur(s) will in fact start a business, the startup proceeds into the seed stage. The effort transforms from being "an idea" or "a group of entrepreneurs thinking together" to a simple form of a company/startup. Typically in this stage, the activities entrepreneurs are involved in include identifying an initial vision, basic business planning, initial technical designs, efforts to identify potential clients, and dividing roles and responsibilities among team members. There are also other activities that might require funding, such as buying hardware components, developing a prototype, conducting market research, or hiring a freelancer to help with a needed task. Bootstrapping is the funding option used in the seed phase; entrepreneurs usually do not pursue external sources of funding this early. This is the best moment in a company's lifetime to cover major functionalities by core team members devoting time and effort because they believe in the idea and its potential. The available funding sources for the seed-stage include personal financing, FFF, accelerators, incubators, business plan competitions, angel investors, and crowdfunding. Target investors are more likely to invest in a team that has a more solid understanding of what it wants to do and how to go about doing it.

When a startup is actively pursuing its product or service, and the basic elements are coming together, it evolves from seed-stage to early-stage. If the company manufactures a product, they probably designed and implemented a prototype during the seed-stage. Now the company must implement the end product that will be delivered to the client and be ready for use. If it is an online service, this is when the word "beta" should be removed and the service should be operational for online clients. This stage identifies the startup's core value proposition, as it interacts with clients and offers services or products, and entrepreneurs are able to identify successful aspects and those that require modification. The early-stage is a period of reflection for the startup entrepreneurs. They have to decide whether to redefine the startup's core value proposition, pivot the business in another direction, or figure out more revenue streams as the entrepreneurs learn more about their potential and their financial needs to stay in business.

By the end of this phase, the business may be growing a client base or selling products and realizing some revenues. Profitability is rare in the early-stage. The funding at this stage is typically used for company establishment and stabilization. Funding at the early-stage is often referred to as series A, a company's first significant round of venture capital investment, or series B its second round. This refers to office rent, office supplies, salaries, sales and customer reach, marketing, legalities, phone bills, manufacturing equipment, and so on. At this stage, a startup typically seeks finance from angel investors or crowdfunding, and in some cases FFF might still be contributing, depending on the scope of business. The startup may also attract the interest of venture capital funding, if it is not too small for a fund's target size. The target investor would be seeking a company with a solid understanding of the business, accompanied by market feedback that reflects on revenues.

Not all companies survive the early stage. In a study focused on the premature scaling of startups, Marmer and Dogrultan[6] state that more than 90 percent of startups fail. But the lucky ones who manage to survive and develop move to the growth stage. This is when they become more attractive and more eligible for larger-scale funding. In the growth stage, a company will be breaking even but may not be sustainably profitable. It has figured out its core value proposition and is growing its client base. It is relatively stable and is thinking of how to grow its current business without affecting existing clients. It will typically need capital at this stage to increase capacity to handle growing business needs, grow its sales team to drive in more revenue, expand to new locations, and possibly rebrand or move to more client-friendly premises. A typical funding source for the growth stage is venture capital in addition to potential interest from private equity firms.

The final stage is the maturity stage. At the maturity stage, the business is no longer a startup; it is a company with real market presence. It has a solid client base, it is making revenue, it is profitable, and it is growing. In this phase, the company's main activities include stabilizing the business and dealing with scalability and management issues. The company's peripheral activities that may require funding include expansion, possibly to another country, or adding new business lines. As the company thinks about its future with regard to possible exit scenarios, such as initial public offerings (IPOs) or acquisition, the main funding sources are private equity firms or IPOs. The choice of exit scenario should be planned ahead to ensure an understanding of outcomes and a maximizing of benefits. For example, an IPO brings in public traders as partners, requires the periodic release of audited financial statements, and entails legal responsibilities. Furthermore, the value of the company will also be dependent on stock supply and demand, as well as a good reputation with a wide client base. Acquisition, on the other hand, is likely to mean that the owner is bought out of the company. In this case, the value of the company is determined based on the deal with the acquirer, and that deal also defines the roles and responsibilities of the owner after acquisition. A summary of the different stages and possible funding sources in the current Egyptian market is presented in table 6.2.

Table 6.2 Funding stages and potential funding sources

	Pre-seed	Seed	Early stage	Growth	Maturity
FFF	Yes	Yes	Yes		
Business plan competitions	Yes	Yes			
Incubator/accelerator	Yes	Yes			
Angel investors		Yes	Yes		
Crowdfunding		Yes	Yes		
Venture capitalists			Yes	Yes	
Private equity firms				Yes	Yes

Source: Based on the author's understanding of the current Egyptian market; may change as the market matures or as more firms and organizations enter the market with different mandates.

Having gone over the different stages in a startup company's lifetime, we will take a deeper look at the various funding sources mentioned and at some examples from the Egyptian market.

Sources of Funding
Friends, family, and fools
FFF are usually the first and easiest to convince to fund a startup. They lack the necessary information to make an informed decision but are hoping to find a good investment opportunity and make returns. Additionally, because FFF have personal ties with the entrepreneurs, they believe in them and hope they will succeed. When speaking to FFF, it is usually a matter of selling the dream of what the startup can become rather than providing hard evidence and business plans. Since FFF are

not organizations and may not have an investment track record, it is important to set clear expectations for both parties. The entrepreneur should know whether to expect funding only or additional support in the form of mentorship, business connections, or sales support. The entrepreneur should also make sure to communicate to FFF what is expected from them as well as the accompanied risks, expected growth, and future plans and scenarios, whether favorable or not.

Business plan competitions

Many institutions and nongovernmental organizations (NGOs) regularly conduct events and activities to support the entrepreneurial ecosystem. Among the most common are business plan competitions. While some business plan competitions offer non-financial prizes, the majority offer a financial prize that can act as a source of funding for startups. Business plan competitions are perfect for companies in the pre-seed stage as they offer a focus on business education and helping entrepreneurs formulate business plans. Mentors usually attend competitions and offer a valuable opportunity to solicit experience and advice.

An example of an NGO hosting business plan competitions is Injaz Egypt. Injaz is a nonprofit organization that acts as a volunteer-based education program with a mission to inspire, empower, and prepare young Egyptians to join the job market as qualified employees and entrepreneurs. Injaz Egypt is part of a regional network called Injaz al-Arab, which harnesses the mentorship of Arab business leaders to cultivate a culture of entrepreneurialism and business innovation among Arab youth throughout fourteen countries in the MENA region. They hold an annual business plan competition, "Startup Egypt," and offer sponsored prizes that can reach LE100,000 (USD15,000).[7] Technology and non-technology related ideas are eligible for consideration. As per their website, Injaz has reached out to 406,000 students; 4,000 volunteers; 358 public schools; 16 universities; 25 governorates; and 31 registered companies.

Incubators/accelerators

Accelerators aim to identify, engage, and support entrepreneurs who have ideas with growth potential. Accelerators periodically announce that they are accepting entrepreneurs and then apply their own selection criteria for admission. Before enrolling in the program, the entrepreneur should have a startup that exists as a legal entity, because accelerators take a percentage of equity in exchange for services provided.

Acceleration time varies, but it is typically three to six months. During this period a number of services are offered, including office space, business education, mentorship, and cash to cover basic expenses. On completion, the accelerator provides exposure to various types of investors. Incubators and accelerators are generally similar in how they alleviate some burdens for the entrepreneur, and the two terms are used interchangeably in many cases. The main difference between them is in their methodology of mentorship, as well as the cash investment and the stake they take in each venture. One key difference in the case of Egypt is that many incubators do not take equity in the startup in return for incubation, while accelerators take equity. Accelerators are more likely to offer mentorship and training as well as seed funding and exposure during a demo day. A demo day is an event held to showcase potential startups to potential investors and allows for the forging of potentially important business connections.

An example of an accelerator in Egypt is Flat6Labs, launched in 2011. Their current model involves holding two cycles a year, during which they select potential startups. The startups then join a boot camp for a final selection to be eligible for the program. Those admitted are offered USD10,000–15,000 and four months of acceleration in return for an average of 10–15 percent of the startup's equity.[8] Flat6Labs accepts only technology or technology-enabled startups. As per their website, they have already accelerated fifty-seven companies and have presence in three locations: Cairo, Jeddah, and Abu Dhabi.

Angel investors

The Organization for Economic Cooperation and Development (OECD) provides many definitions of angel investors. One of their definitions that is relevant here is:

> A business angel is an individual investor (qualified as defined by some national regulations) that invests directly (or through their personal holding) their own money predominantly in seed or startup companies with no family relationships. Business angels (BAs) make their own (final) investment decisions and are financially independent, meaning that a possible total loss of their business angel investments will not significantly change the economic situation of their assets. BAs invest with a medium- to long-term set timeframe and are ready to provide, on top of their individual investment, follow-up strategic support to entrepreneurs from investment to exit.[9]

One of the most popular angel investor networks in Egypt is Cairo Angels. In Cairo Angels, each member makes his or her individual investment decision, but they collaborate in due diligence. Collectively, they make equity investments in the range of LE250,000 to LE1 million per company, and can syndicate larger investments to partner organizations.[10] The network's strategy is to not take majority stakes. Since its inception in early 2012, Cairo Angels has invested more than LE5 million in eight startups with binding offers being made almost on a quarterly basis. The goal of this angel investor network is to triple their funding base and deal flow by the end of 2015.

Crowdfunding

Crowdfunding has grown worldwide since 2009, and in Egypt since the end of 2012. Crowdfunding is based on the idea of utilizing the power of the crowd to have an impact that extends far beyond that of a single person or institution. Crowdfunding brings together the efforts, resources, and money of a group of people to make a project or company succeed. Instead of trying to raise a lot of money from a single investor, many smaller investors are involved, thus increasing the chances of receiving funds and also diluting the investment risk among those involved. Moreover, crowdfunding offers an important advantage by involving potential future clients in validation. If a significant number of people are willing to invest in a business, this means that it makes sense to them as users/ clients, which is a very important validation for a startup.

Crowdfunding can take place through a variety of models. A useful model for entrepreneurs and startups in the funding gap is equity-based crowdfunding. This is a novel yet growing solution. A startup is typically posted on a crowdfunding platform to raise money for a predetermined amount of equity, and the platform itself is rewarded only on successful transactions.

The only equity-based crowdfunding solution in Egypt at this time is Shekra, cofounded by the author of this chapter. Shekra, short for Sharek/Fekra, which translates into "share an idea" in Arabic, links potential startups with a closed network of investors in a partnership model. Shekra's model is based on three phases. The first is the pre-funding phase, where basic business education and help in business plan writing is offered as well as collaborative work on valuation and fair equity stake. The second phase is the actual crowdfunding stage, where the startup is posted on a portal for investors to consider funding. Upon successful completion of funding, the third and final phase begins. This involves the legal structuring of the company to include new shareholders, in addition to mentorship and monitoring activities to ensure the business is on track.

Shekra guarantees that all business is conducted in a manner compliant with sharia, which is Islamic law. Also, instead of taking a percentage of the money raised, as crowdfunding portals tend to do worldwide, this platform takes in-kind equity in each funded startup, thus aligning its interests with both the entrepreneurs and investors. Shekra accepts both technology and non-technology startups, removing the sector limitation that some entrepreneurs face.[11]

Venture capital

Venture capital is defined as providing "long-term, committed share capital, to help unquoted companies grow and succeed."[12] Venture capital funds own equity in the companies they invest in and diversify their portfolios by investing in several companies. An example from the Egyptian market is Ideavelopers, a venture capital firm that helps innovative startups grow to become successful companies. Ideavelopers is a subsidiary of EFG-Hermes Private Equity. The firm manages the Technology Development Fund, a usd50 million fund focused on early-stage technology companies. Their mission is to invest in and support startups with significant potential.[13] They not only provide risk capital to their portfolio companies but also support them with strategic advice and industry access. When evaluating a startup, Ideavelopers looks for strong teams with differentiated offerings that target sizable markets and can consequently realize attractive financial returns. They have so far invested in seventeen technology companies.

Private equity

One of the leading independent wealth and asset managers in Europe, founded in 1805, the Pictet Group, defines private equity as follows:

> Private equity capital is equity capital that is not quoted on a public exchange. Private equity investors or funds make investments directly into private companies or conduct buyouts of public companies that result in a delisting of public equity. Capital for private equity is raised from retail and institutional investors and can be used to fund startups (venture capital), make acquisitions (growth equity, buyout), or to strengthen a balance sheet (special situations). Such investments are commonly made by private equity firms, venture capital firms, or 'angel investors.'[14]

A good example of a private equity firm in the Egyptian market is Citadel Capital, which recently changed its English name to Qalaa Holdings. The holding company comprises ten subsidiaries. Citadel Capital, CCAP. CA on the Egyptian Exchange, is a leading investment company in Africa and the Middle East. The firm builds businesses in five core industries: energy, transportation, agri-foods, mining, and cement.[15]

Development organizations

One of the most important catalysts for any ecosystem is development organizations. While they might not always provide direct funding, such organizations provide indirect funding in the form of capacity building or funding for business plan competitions. One example is USAID Egypt's Competitiveness Program (ECP), which supports 'Startup Weekend Egypt.' This event gathers entrepreneurs to advance ideas with the help of mentors over a weekend. Other examples include the World Bank Group, which is working with the Social Fund for Development, a governmental organization that provides credit facilities at competitive interest rates to small entrepreneurs through numerous banking and commercial institutions or through direct lending.

Which Source to Approach

Finally, after verifying investment readiness and obtaining sufficient information to choose the correct funding option, entrepreneurs must make sure to select the most suitable source of funding. This choice can be expanded to areas beyond funding, based on the needs of the company at its specific stage and the team's understanding of their business and how they intend to grow it, as highlighted previously in this chapter. MENA Private Equity Association[16] points out these areas to consider when choosing a funder:

- In which stage of the company's development does the firm usually invest?
- What industry sectors do they focus on?
- What is the firm's geographical focus and does it match the company's present and future plans?
- What equity stake do they typically seek (minority or majority)?
- What is their typical minimum and maximum ticket size (investment size), and what range is their sweet spot?
- What board representation do they require?

- How involved are they in their portfolio management post-acquisition?
- What networks and relationships do they have that can benefit the business?
- What is their track record in similar industries and geographies?
- How big is their team on the ground?

Entrepreneurs starting their own venture face a challenging task. It takes a lot of dedication and involves acquiring knowledge in unfamiliar areas. Entrepreneurs must also always have their eyes on the market. This involves soliciting feedback from professionals and clients in order to determine whether the business is on track or whether the business model requires changes.

When pursuing funding, entrepreneurs are pitching their business case to investors and answering questions to convince them to bet on their team and idea and be willing to share the associated risk. Assessing the business from all angles and making sure it is sellable is essential. It is healthy to periodically go through the exercise of assessing the business from an investor's point of view to determine value as well as value added. This chapter serves as a guide to those steps and how to think about each: understanding investment readiness and which stage the company is in, going over the various sources of funding and studying each of them, and finally choosing which source to approach and being ready to grow into a successful company.

Notes

1 Silatech and Gallup Inc., *The Silatech Index: Voices of Young Arabs* (Doha, Qatar; Silatech and Gallup Inc., 2010).
2 Laura Goldman, Rasesh Mohan, and Andrew Stern, *Investor and Funder Guide to the Agricultural Social Lending Sector*, The Initiative for Smallholder Finance, 2014.
3 'Bootstrapping' refers to using currently available resources, personal financing, or generated revenues to the maximum before turning to external resources, that is, pursuing investment.
4 "Startup Check List," RBC Royal Bank, http://www.rbcroyalbank.com/sme/getting-ready/startup.html.
5 "Angel Investment Checklist," New York Angels, http://www.newyorkangels.com/angel-investment.html.
6 Max Marmer and Ertan Dogrultan, *Startup Genome Report Extra* (UC Berkeley: & Stanford Startup Genome Project, 2011).
7 Injaz Egypt, "Injaz Egypt," http://injaz-egypt.org/.
8 Flat6Labs, "Flat6Labs," http://www.flat6labs.com/.
9 OECD, *Financing High-Growth Firms: The Role of Angel Investors* (Paris: OECD Publishing, 2011), 29, http://dx.doi.org/10.1787/9789264118782–en.

10   Cairo Angels, "Cairo Angels: A Forum for Investors to Find High Potential Start Ups," http://www.cairoangels.com/
11   Shekra, "Shekra: Crowd-funding Solutions," http://www.shekra.com/en/.
12   Intertrade Ireland and Irish Venture Capital Association (IVCA), *Guide to Venture Capital*, Ireland: Intertrade Ireland and Irish Venture Capital Association (IVCA), 5th ed., 2012), 6.
13   Ideavelopers, "Ideavelopers: From Ideas to Success," http://www.ideavelopers.com/.
14   According to PICTET Alternative Advisors SA, *An Introduction to Private Equity*, Geneva: The Pictet Group, 2014), 5.
15   Qalaa Holdings, "Qalaa Holdings: An African Leader in Infrastructure and Industry," http://www.qalaaholdings.com/.
16   MENA Private Equity Association, the VC Taskforce Team, and industry professionals, *A Guide to Venture Capital in the Middle East and North Africa* (The Veeceepreneur, 2013), 8.

Bibliography
Cairo Angels. "Cairo Angels: A Forum for Investors to Find High Potential Start Ups." http://www.cairoangels.com/.
Flat6Labs. "Flat6Labs." http://www.flat6labs.com/.
Goldman, Laura, Rasesh Mohan, and Andrew Stern. *Investor and Funder Guide to the Agricultural Social Lending Sector.* New York – Washington: The Initiative for Smallholder Finance, 2014.
Ideavelopers. "Ideavelopers: From Ideas to Success." http://www.ideavelopers.com/.
Injaz Egypt. "Injaz Egypt." http://injaz-egypt.org/.
Intertrade Ireland and Irish Venture Capital Association (IVCA). *Guide to Venture Capital.* Ireland: Intertrade Ireland and Irish Venture Capital Association (IVCA), 5th ed., 2012.
Marmer, Max, and Ertan Dogrultan. *Startup Genome Report Extra.* UC Berkeley & Stanford: Startup Genome Project, 2011.
MENA Private Equity Association, the VC Taskforce Team, and industry professionals. *A Guide to Venture Capital in the Middle East and North Africa.* The Veeceepreneur, 2013.
New York Angels. "Angel Investment Checklist." http://www.newyorkangels.com/angel-investment.html.
OECD. *Financing High-Growth Firms: The Role of Angel Investors.* Paris: OECD Publishing, 2011. http://dx.doi.org/10.1787/9789264118782–en
PICTET Alternative Advisors SA. *An Introduction to Private Equity.* (Geneva), The Pictet Group, 2014.
Qalaa Holdings. "Qalaa Holdings: An African Leader in Infrastructure and Industry." http://www.qalaaholdings.com/.
RBC Royal Bank. "Startup Check List." http://www.rbcroyalbank.com/sme/getting-ready/startup.html.
Shekra. "Shekra: Crowd-funding Solutions." http://www.shekra.com/en/.
Silatech and Gallup Inc. *The Silatech Index: Voices of Young Arabs.* Silatech and Gallup Inc., 2010.

7. Building a University-centered Entrepreneurship Ecosystem: A Case Study of the American University in Cairo

Ayman Ismail

Introduction

In this chapter, I analyze the entrepreneurship ecosystem at the American University in Cairo (AUC) School of Business, providing a case study of how universities may contribute to building an entrepreneurial ecosystem in an emerging economy. The chapter also identifies a number of best practices and lessons learned for university-centered entrepreneurship programs and startup accelerators.

In 2010, the AUC School of Business established the Entrepreneurship and Innovation Program (EIP) with a broad mission to spread awareness and educate young people, both at AUC and beyond the university, about entrepreneurship and innovation.[1] I examine how EIP has contributed to building the entrepreneurship ecosystem in Egypt in six key areas: entrepreneurs, ideas, networks, mentors, funding, and startups. This framework has guided the activities, programs, and performance evaluation of EIP. In each of these areas, EIP has implemented a number of activities, programs, and partnerships. The objectives are raising awareness, facilitating startup team formation, creating new innovative business models through ideation workshops and business plan competitions, facilitating the creation of startup ventures through startup weekends, building startups' business plans, and connecting startups with potential mentors, funders, team members, and customers. Based on the feedback from EIP activities, a clear need to provide in-depth services to a small number of serious startups was evident.

In 2013, AUC established the AUC Venture Lab, a university-based accelerator/incubator, with a mission to identify, incubate, mentor, connect, and support those who possess the talent and desire to become entrepreneurs, facilitating their success in launching startup ventures.

There is a general consensus that entrepreneurs play a key part in advancing a country's economy,[2] which implies an important role for universities in fostering and promoting an ecosystem for innovation and entrepreneurship.[3] Harvard Business School was the first university to offer courses on entrepreneurship in 1948.[4]

Yet Kirby[5] argues that "it was only in the last two decades of the twentieth century that any considerable attention was paid by academia to the role of higher education in the creation of graduate entrepreneurs."[6] In 2000, the United Kingdom named entrepreneurial development as a strategic goal for British universities.[7] In the broader European region, entrepreneurship began evolving as a significant part of the curriculum ten years ago, for example at the Universidad Autonoma de Madrid in Spain, Delft University of Technology in the Netherlands, and the GEA College of Entrepreneurship in Slovenia,[8] with some minor exceptions of universities that started earlier.[9]

The variety of approaches toward entrepreneurship—for example high-growth, innovation-driven entrepreneurship versus small and medium enterprises—is an important factor that shapes the programs offered by different universities. In the United States, entrepreneurship commonly refers to innovation-driven and growth-oriented ventures or companies.[10] Universities there have a critical role in connecting academics, researchers, and students with the business world by developing networks with entrepreneurs, businesspeople, venture capitalists, and angel investors.[11]

In Europe, entrepreneurship has traditionally been synonymous with small and medium enterprises (SMEs), which leads most programs to concentrate on teaching functional management skills for small business owners.[12] The link between the university and the outside private business sector is minimal, and there is a tendency to encourage students to secure future jobs rather than become entrepreneurs.[13] Universities in Germany, conversely, do tend to take a more pro-entrepreneurship approach,[14] in the American sense. These institutions often build internal support structures of professors, departments, and other support services, including "entrepreneurship centers and technology transfer units," which offer consultation and networks to future entrepreneurs.[15] Similarly, legal regulations in France introduced in 1999–2000 encourage collaboration between universities and companies around new technologies and fostering an entrepreneurial mindset among students.[16]

In the MENA region, similar to most of the global south, numerous programs to support SMEs have been implemented over the past two decades, with the main objective of generating employment, especially among young people.[17] These programs were supported by international donors, national governments, and financial institutions. Programs to support high-growth, innovation-based entrepreneurship have only become visible in the past few years, however.[18] Over the past five years, a number of accelerators/incubators,[19] venture capital funds,[20] angel investor networks,[21] and other support programs have been introduced, mostly by visionary entrepreneurs.

One of the ways that universities support and promote entrepreneurship and innovation is by creating and hosting business incubators and accelerators. Business incubators provide services such as managerial assistance, counseling, and networking.[22] They also offer tangible support such as equipment and office space, and resources such as research labs and seminars.[23] The goal of university-based incubators varies between (a) facilitating the startup of new companies, increasing their survival rate and growth, and, more generally, by training entrepreneurs, and (b) stimulating firms involved in emerging technologies or the commercialization (or transfer) of research done in universities, research institutions and firms."[24]

Entrepreneurship and Innovation in Egypt

According to the Egypt Entrepreneurship Report 2010 by the Global Entrepreneurship Monitor (GEM),[25] in Egypt the total entrepreneurial activity (TEA) rate—which measures the percentage of the population (18–64 years old) either actively trying to start a business or already owning and managing a business—was 7 percent in 2010 (ranked 37 among the 59 TEA-measured countries). The majority of the young entrepreneurs who chose to pursue entrepreneurship indicated that they had done so due to a lack of better employment opportunities. This low entrepreneurial activity took place at a time of high economic growth: Egypt's GDP in 2008 witnessed a growth rate of 7.2 percent, primarily driven by the private sector, which generated 68 percent of the total growth.[26] Having low entrepreneurial activity indicators at times of high economic growth indicates that most of the growth was driven by 'older' companies rather than new startups.

In response to the low rate of entrepreneurial activity, the Egyptian government has established several strategies to encourage entrepreneurship. These include training programs, financing opportunities, and technical support.[27] Rules and regulations have also seen some shifts. Regarding the ease of doing business, Egypt was considered one of the top global reformers when it came to simplification of administrative work in 2007.[28] For example, the creation of "one-stop-shops" to consolidate government services in one location has helped streamline and facilitate the process of starting up a new business.[29] Most of these reforms have targeted large investors and corporations, though, rather than small startup companies.

The government has also supported entrepreneurship, albeit mostly SMEs rather than high-growth, innovative entrepreneurship, through financial opportunities. Public banks such as the National Bank of Egypt, Banque du Caire, Banque Misr, and the Bank of Alexandria have created departments to address the particular needs of SMEs.[30] Additionally, the Social Fund for Development (SFD) and the Industrial Modernization Center (IMC), both quasi-governmental entities, have created SME support programs.[31]

Looking at the entrepreneurial ecosystem as a whole, El Dahshan, Tolba, and Badreldin identified some of the most active organizations in Egypt that support entrepreneurship.[32] These are the Information Technology Industry Development Agency (ITIDA), the SME Development Unit at the Ministry of Finance, the Middle East Council for Small Business and Entrepreneurship (MCSBE), Nahdet El Mahrousa, Ashoka, Entrepreneurs Business Forum (EBF), Endeavor, Alashanek ya Baladi (AYB-SD), the Egyptian Junior Business Association (EJB), the Women Entrepreneurship and Leadership (WEL) Program at AUC, the Center for Entrepreneurship at Cairo University, and the European Training Foundation (ETF). The demographics of Egypt provide an incredible platform to grow such a space with over 58% of the population under the age of 25 and very much eager to explore the world of the private sector.[33]

Despite these efforts by the government and international organizations, entrepreneurial activity in Egypt remains low. Additionally, entrepreneurship education in Egypt is generally considered weak. In the Global Entrepreneurship Monitor (GEM), Hattab[34] illustrates this by demonstrating how school and university students typically do not obtain the necessary business education that would allow them to start and operate a business.[35] Nonetheless, experts claim that the vocational training of some young Egyptians assists them to a great extent in starting businesses.[36]

Findings from background research into startups in Egypt demonstrated that there is a substantial need in two areas: first, a need for promoting entrepreneurship among young people and providing them with tools to generate ideas and start their ventures, and second, a need for support services, especially for early-stage entrepreneurs as they build their startups.[37] In light of these needs, I examine how a university can contribute to building an entrepreneurial ecosystem in an emerging economy, taking the Entrepreneurship and Innovation Program at AUC as a case study.

The Entrepreneurship and Innovation Program (EIP)

In response to the market need for support services for entrepreneurs, the American University in Cairo (AUC) School of Business launched the Entrepreneurship and Innovation Program (EIP) in October 2010. The aim was "to create an environment that fosters the development of principled and innovative business leaders and entrepreneurs who can make a difference."[38] The mission of EIP is to promote entrepreneurship among young Egyptians, from all universities. In collaboration with different stakeholders within AUC and partners in the business community, and through seminars, workshops, networking events, mentorships, business boot camps, and business plan competitions, the program aims to educate, train, and inspire students as to what entrepreneurship is. EIP thus helps aspiring entrepreneurs generate ideas for businesses and then connects the most viable startups to the needed resources and networks.

In the early stages of building the EIP ecosystem, discussions and meetings were held with various stakeholders sharing a common entrepreneurial passion. These conversations included faculty, students, alumni, and business leaders with specific interest in the area. To formalize these discussions, the school established the Entrepreneurship and Innovation Program Council to serve as an advisory board for the program. The EIP Council had among its members entrepreneurs, academics, policy makers, and business leaders with an interest in entrepreneurship. Over time, they became judges and mentors in competitions, acted as angel investors, and provided advice in designing various EIP programs and activities. EIP also established a network of practitioners, business executives, and academics interested in mentoring and coaching entrepreneurs at various stages of their startup journeys.

EIP takes a comprehensive ecosystem approach in designing its action framework, focusing its activities on six key areas: entrepreneurs, ideas, networks, mentors, funding, and startup ventures. The program focuses heavily on partnerships with other organizations to implement its activities.

The initial focus area of the framework, entrepreneurs, is implemented by raising awareness about entrepreneurship among the participants, who vary in education, socioeconomic background, and age group. Additionally, this stage acts as a catalyst for startup team formation and exposes the entrepreneurs to the venture process and the ecosystem.

The second area, ideas, revolves around generating attractive ideas, conceptualizing business opportunities, and finally developing business plans. Attractive ideas are those that respond to the Egyptian market's needs. Leadership panels, partnerships with incubators, and summer camps are the major activities of this stage.

Network creation is the third area of EIP's framework. Through the collaboration of twenty-eight universities, companies, and international institutions, and through the involvement of business executives from a variety of sectors, participants are exposed to real-life examples of entrepreneurship. Meetings and discussions are carried out between like-minded entrepreneurs, industry experts, and local leaders.

The mentorship level focus revolves around coaching and mentoring potential entrepreneurs through the development of their business plans and launch of their startups. Furthermore, mentors provide internships in startups. The mentoring process is done through the AUC mentors network, and supported by faculty advice, workshops, and training.

Following this phase, entrepreneurs are encouraged to seek funds. The program connects entrepreneurs to venture capitalists, angel investors, and potential investment partners. EIP also offers financial awards through startup competitions.

Last, in an attempt to connect startup ventures to the market, some entrepreneurs are admitted to business incubators and others are assisted in promoting their ideas to the market. This is done by connecting startups to incubators and accelerators, supporting incubated startups in partner organizations, or providing visibility and access to startups.

Working with young entrepreneurs highlighted the need for providing additional in-depth services to serious early-stage entrepreneurs/startups as they worked through their business modeling and planning, fundraising, and setting up their operations and partnerships. These services are best provided to a smaller number of startups through an acceleration/incubation program. This provided the motivation to expand the scope of the entrepreneurship ecosystem at AUC by establishing the AUC Venture Lab.

AUC Venture Lab

In 2013, the American University in Cairo established the AUC Venture Lab, a university-based incubator to provide in-depth support services for a small number of serious entrepreneurs and their startups.[39] Findings from background research conducted on startups in Egypt demonstrated that there is a huge "white space" when it comes to the provision of services in the market.[40] Many startups were in need of services that could be easily offered by a university-based incubator, such as mentorship and coaching, networking and connections, and access to university facilities, faculty, and students.

The business model of the Venture Lab was shaped by research conducted to study other university-based incubators around the world,[41] which provided insights into their various business models. Globally, for example, universities tended to select companies that matched their own internal competencies. Many indicated a significant interest in technology, and the time period between affiliation and incubation ranged from several months to several years. Many also offered multiple incubator types and stages, allowing a diversity of entrepreneurs to enter their programs. Compared to those in the United States, emerging market programs tended to offer longer incubation periods (up to eighteen months), and university faculty tended to demonstrate a more intimate, one-on-one relationship with the incubated entrepreneurs. Short-term incubators were a double-edged sword: on the one hand, they may push entrepreneurs to get their products to market more quickly, but they may also rush products that need more time to develop.[42]

In addition to all the above-mentioned EIP activities and programs, the Venture Lab offers a few serious entrepreneurs a variety of services aimed at assisting the startup process, increasing business survival rates, and providing avenues for access to funding from angel investors, venture capitalists, or other sources. The Venture Lab utilizes the university's capabilities (knowledge, faculty, staff, facilities, space, brand name, and services) to help companies with strong growth potential launch successfully.

The Venture Lab provides two cycles of incubation each year, each running for four months. Five to eight startup teams are selected through a competitive process to join the program. Entrepreneurs from all over Egypt are accepted, even if they are not affiliated with AUC, as this encourages diversity and meritocracy. During the acceleration stage, the Venture Lab provides a small seed fund to each of the accepted startups, along with a set of services, which are listed in table 7.1.

The Venture Lab targets startups after the idea stage and prior to entering the market. This phase fits well with the university-based incubator model, as at this point entrepreneurs have a general idea of what their product or service will look like as well as a prototype or pilot but still require a significant amount of technical and business support. Startups entering the program must have or be working on a prototype, pilot, or proof of concept for their product or service. The Venture Lab is sector-agnostic but requires that entrepreneurs have an innovative approach to solving or filling existing demand with a unique value proposition. Through its partners, the Venture Lab has reached out to students in all seventeen public universities in Egypt to ensure a diverse and interesting pool of applicants.

Startups go through a rigid two-month selection process that includes a detailed application, initial presentations, and finally the pitching of ideas to a panel of seasoned entrepreneurs and investors. The selection criteria cover three main areas: first, the business opportunity or idea must be original, impact on a problem, fill a market gap, be innovative, and fit with AUC Venture Lab service offerings; second, the business must have passed the idea stage and have developed a prototype, and the viability of its revenue model and cash burn rate will be examined; and third, the entrepreneur must have commitment to the business, managerial capabilities, and an acceptance of feedback.

The selection process is designed to add value to the entrepreneurs even if they are not selected for incubation. Before final presentations, all companies are required to attend an interactive training program that focuses on building business skills. This also enables the Venture Lab to work closely with each entrepreneur and evaluate his or her talents, abilities, and motivation. The program is designed and led by AUC faculty, business practitioners, and executives selected from our mentor network.

Based on the selection process, startups are admitted into a four-month acceleration program. During this period, the Venture Lab educates startups on basic business skills, works with them to finalize business models and develop functioning prototypes of products or services, and connects them to business leaders and mentors. A "startup boot camp" training program explores basic business principles, and coaching and mentorship are offered in tandem. Facilities including labs, theaters, and mass communications are offered, as well as workspaces. Entrepreneurs are offered help in recruiting other students, especially interns, to join their projects.

Table 7.1. Services provided by the Venture Lab to incubated startups

Co-working space	Companies are granted access to a shared co-working space on campus during the acceleration cycle. After that time, some may be granted an extension based on need and performance.
Training	A week-long boot camp (training program) is offered at the beginning of the incubation cycle, covering the fundamentals of management as well as specific topics of interest for startups, such as fundraising and idea-pitching.
Mentoring and coaching	Startups are supported through a mentoring program that provides access to advisors and coaches in their respective industries and sectors, based on need.
Access to investors	During the incubation period, the Venture Lab organizes networking events with investors and connects incubated companies with potential investors.
Other services	Other services, which are offered as needed, include access to students for employment and/or market research; access to faculty as mentors and/or consultants; assistance with professional services such as human resources and recruitment, communication, and legal assistance; and access to other AUC facilities upon request and agreement, including AUC engineering and technical labs.

The acceleration phase provides training to students in five areas of business management. First, they are taught the basics of planning a business, including the business model, market, product, and value proposition. They are introduced to project planning tools and taught how to create a business plan. The second aspect involves developing a "product that works." Next, launching that product requires skills in marketing, advertising, and sales. Financial management explores aspects such as equity management, financing, budget, and cash flow management, as well as accounting and taxes. Finally, training on "organizing for growth" helps startups learn how to manage people, as well as organizational values/culture, in an early-stage organization. Upon finishing the four-month cycle, startups are expected to have a finalized business plan, a working product or prototype, and a financial plan.

The coaching program supports the team in areas that don't require deep expertise but rather general management experience. Coaches are selected based on expressed interest as well as matching during events (for example, speed mentoring). Each company is assigned one coach, who is requested to meet frequently. Coaches are generally individuals with relevant work experience and a deep interest in the company's idea/business model. They also should have the capability to pitch and/or defend the idea in front of investors.

Speaking and networking events are held on a regular basis. These are open to outside entrepreneurs, and include pitches, speed mentoring, sharing of success stories, and relevant topic-based talks. In addition, the Venture Lab assists with fundraising by providing access to an angel network and support in negotiating deals.

Promising startups may be offered an additional nine-month incubation period. This primarily involves customized support for the startup, such as workspaces, the use of facilities, and guidance. Entrepreneurs may be advised on human resources and legal support and participate in three- to nine-month mentorship programs.

Thus far, the Venture Lab has completed two acceleration cycles, in which it received over three hundred applicants and incubated eleven startups from diverse sectors, who are described in table 7.2.

Table 7.2. Profile of select AUC Venture Lab startups

XOLOGY	**Axology** deals with the "science of accessories." A product innovation firm, it designs and builds accessories to optimize security and efficiency in cargo handling.
BUS POOLING	**Bus Pooling** is a subscription-based bus service that transports commuters between home and work. After submitting a request, Bus Pooling matches individuals living in the same area, who share the same work location and hours, and supplies a bus and schedule customized to meet these needs.
DoubleVee	**DoubleVee** stands for verification and validation. The first Egyptian company to specialize in software testing, DoubleVee performs functional, performance, operational, and security testing services according to international best practices and standards.
Madad	**Madad** is an online platform that provides a directory of sustainable development projects in Egypt. The platform offers potential donors a gateway to financially support appealing projects, track the use of their donations, and follow the execution of the projects.

	Tatweer is a social business whose mission is to upgrade and develop under-utilized trade channels, starting with the street kiosk.
	Alkottab is an "edutainment" games studio. Alkottab believes that games are not just for entertainment but can promote important educational values and change perceptions of how to approach and solve problems.
	En2ly is a web and mobile service connecting Egypt's fragmented freight transportation industry. In Egypt, individuals own 90 percent of an estimated one million freight trucks. It is estimated that half of these trucks are routinely under-utilized, and often carry no shipments when returning from a delivery destination, wasting fuel and contributing to environmental decay.
	Kashef Labs is developing a ground-penetrating radar capable of detecting the many landmines left by the Axis forces in World War II in Egypt's borderlands. Using an unmanned aerial surveillance tool that is lightweight and utilizes minimal power, the radar flies one meter above desert rock and sand to scan the ground.
	Mubser develops wearable technology to aid visually impaired people in their everyday lives. Mubser's pilot product, Sensify, coordinates the user's smartphone or Mubser pocket computer to detect obstacles and notify the user through vibrations on a bracelet and a Bluetooth headset.
	El Shahbander helps textile workers in Egypt communicate with one another, introducing them to manufacturers and importers through a website and mobile application. El Shahbander is the first specialized community in the Egyptian textile industry, bringing global trends to the local market.
	Smart News is a multiplatform content application that provides localized news organized by country, category, and interest. Users can select the source of their news, as well as the topics, which include national and international news coverage, sports, fashion, and technology.

Conclusions and Lessons Learned

The AUC Venture Lab has developed a number of key performance indicators (KPIs) to measure its performance and impact. The main KPI is the number of successful startups the Venture Lab will create; additional indicators include the number of entrepreneurs trained, the number of startup ventures incubated and supported by the AUC Venture Lab services, and the percentage of startups that access funding (and the amount of that funding) as a result of being incubated. Mentors being able to link entrepreneurs with business executives and experienced entrepreneurs, and the creation of partnerships between startups, business, government, and educational institutions, are additional indicators of success.

Based on interviews with the EIP and Venture Lab teams, program beneficiaries, and entrepreneurs, I identified four key lessons learned. First, it is very important to use a partnership model to collaborate with other players in the ecosystem. Partnerships allow for synergies among the different players and strengthen the ecosystem. Second, it is critical to build on the assets and strengths of AUC as a university in designing the program and incubator. For example, it is crucial to focus on areas where there is interaction between the incubated startups, students, alumni, and faculty, in addition to establishing linkages between startups and AUC facilities, labs, and services. Third, I would advocate the use of an experimental and gradual approach, such as starting with a small number of startups and focusing on their growth. This allows the team to learn and adapt its business model and services to the local needs and context. Finally, the creation of local stakeholders within and outside the university is an important factor for the success of the program. Partnerships form a base of supporters and stakeholders who care about the program and seek its success, which is critical in overcoming many of the challenges and risks associated with operating in emerging market environments.

Notes

1   *Entrepreneurship and Innovation Program (EIP)* (Cairo: American University in Cairo Press, 2010).

2   AfDB, *Egypt Private Sector Country Profile* (African Development Bank, 2009); Hala Hattab, *Global Entrepreneurship Monitor: Egypt Entrepreneurship Report, 2010* (Cairo: The Industrial Modernization Centre, 2010); Karen Wilson, *Entrepreneurship Education in Europe* (Brussels: European Foundation for Entrepreneurship Research, 2008); Mohamed El Dahshan, Ahmed Tolba, and Tamer Badreldin, *Enabling Entrepreneurship in Egypt: Towards a Sustainable Dynamic Model* (Alexandria: Entrepreneurship Business Forum, Egypt, 2010).

3   Wilson, *Entrepreneurship Education in Europe*.

4   Wilson, *Entrepreneurship Education in Europe.*

5   David Kirby is a British researcher and professor of entrepreneurship. He was the founding dean of the British University in Egypt and the former dean of the business school at Middlesex University in the UK.

6   David Kirby, "Entrepreneurship Education: Can Business Schools Meet the Challenge?" *Education + Training* 46 (2004): 359.

7   UK Universities, *A Forward Look: Highlights of Our Corporate Plan 2001–2004* (London England: Universities UK, 2000), http://www.universitiesuk.ac.uk/ Publications/Documents/CorpPlan2.pdf

8   NIRAS, *Survey of Entrepreneurship Education in Higher Education in Europe* (Allerød: NIRAS Consultants, 2008).

9   Wilson, *Entrepreneurship Education in Europe.*

10 Wilson, *Entrepreneurship Education in Europe.*

11 Wilson, *Entrepreneurship Education in Europe.*

12 Wilson, *Entrepreneurship Education in Europe.*

13 Wilson, *Entrepreneurship Education in Europe.*

14 Andrea-Rosalinde Hofer et al., *From Strategy to Practice in University Entrepreneurship Support: Strengthening Entrepreneurship and Local Economic Development in Eastern Germany: Youth, Entrepreneurship and Innovation* (Germany: OECD Publishing, 2010).

15 Hofer et al., *From Strategy to Practice.*

16 Christophe Schmitt, *Entrepreneurship and University: Reflections on the Role of the University Incubators* (Belfort, France: 12th International Conference on Management of Technology, 2003).

17 Qamar Saleem, *Overcoming Constraints to SME Development in MENA Countries and Enhancing Access to Finance* (Washington, DC: International Finance Corporation/The World Bank Group, 2013).

18 Sherif Kamel, "Entrepreneurial Uprising," *BizEd* (November/December 2012): 46–47.

19 See, for example, Flat6Labs, Oasis 500, AUC Venture Lab, and Nahdet El Mahrousa.

20 See, for example, Ideavelopers, N2V, and Sawari Ventures.

21 See, for example, Cairo Angels.

22 Christophe Schmitt, *Entrepreneurship and University.*

23 Hanoku Bathula, Manisha Karia, and Malcolm Abbott, "The Role of University-based Incubators in Emerging Economies," Working Paper no. 22 (2011).

24 Anna Bergek and Charlotte Norrman, "Incubator Best Practice: A Framework," *Technovation* 28 (2008): 20–28.

25 Hattab, *Global Entrepreneurship Monitor.*

26 El Dahshan, Tolba, and Badreldin, *Enabling Entrepreneurship in Egypt.*

27 Hattab, *Global Entrepreneurship Monitor.*

28 OECD, *Overcoming Barriers to Administrative Simplification Strategies: Guidance for Policy Makers* (OECD Publishing, 2009).

29 IBRD, *Doing Business 2012: Doing Business in a More Transparent World* (Washington, DC: The International Bank for Reconstruction and Development/The World Bank Group, 2011).

30 AfDB, *Egypt Private Sector Country Profile.*

31 AfDB, *Egypt Private Sector Country Profile.*

32 El Dahshan, Tolba, and Badreldin, *Enabling Entrepreneurship in Egypt.*

33 Kamel, "Entrepreneurial Uprising," *BizEd* (November/December 2012): 46–47.
34 Hala Hattab is a lecturer in entrepreneurship at the British University in Egypt and the author and program manager for the Global Entrepreneurship Monitor (GEM) report in Egypt.
35 Hattab, *Global Entrepreneurship Monitor.*
36 Hattab, *Global Entrepreneurship Monitor*; Munther Masri, Mohamed Jemni et al., *Entrepreneurship Education in the Arab States* (Beirut: UNESCO, 2010).
37 Ayman Ismail and Rawiah Abdallah, "AUC Venture Lab Business Model," unpublished internal report (Cairo: American University in Cairo, 2013).
38 Entrepreneurship and Innovation Program," American University in Cairo, http://www.aucegypt.edu/business/eip/Pages/default.aspx.
39 AUC Venture Lab, http://www.aucegypt.edu/Business/eip/Pages/Venture%20 Lab.aspx.
40 Ismail and Abdallah, "AUC Venture Lab Business Model."
41 Ismail and Abdallah, "AUC Venture Lab Business Model;" Ayman Ismail and Sherif Shabana, "Benchmarking University-based Incubators," unpublished internal report (Cairo: American University in Cairo, 2013).
42 Ismail and Shabana, "Benchmarking University-based Incubators."

Bibliography

AfDB. *Egypt Private Sector Country Profile.* African Development Bank, 2009.
American University in Cairo. "Entrepreneurship and Innovation Program." http://www.aucegypt.edu/business/eip/Pages/default.aspx
Bathula, Hanoku, Manisha Karia, and Malcolm Abbott. "The Role of University-based Incubators in Emerging Economies." Working Paper no. 22. 2011.
Bergek, Anna, and Charlotte Norrman. "Incubator Best Practice: A Framework." *Technovation* 28 (2008): 20–28.
El Dahshan, Mohamed, Ahmed Tolba, and Tamer Badreldin. *Enabling Entrepreneurship in Egypt: Towards a Sustainable Dynamic Model.* Alexandria: Entrepreneurship Business Forum, Egypt, 2010.
Entrepreneurship and Innovation Program (EIP). Cairo: American University in Cairo, 2010.
Hattab, Hala. *Global Entrepreneurship Monitor: Egypt Entrepreneurship Report, 2010.* Cairo: The Industrial Modernization Centre, 2010.
Hofer, Andrea-Rosalinde, Jonathan Potter, Alain Fayolle, Magnus Gulbrandsen, Paul Hannon, Rebecca Harding, Åsa Lindholm Dahlstrand, and Phillip H. Phan. *From Strategy to Practice in University Entrepreneurship Support: Strengthening Entrepreneurship and Local Economic Development in Eastern Germany: Youth, Entrepreneurship and Innovation.* Paris: OECD Publishing, 2010.
IBRD. *Doing Business 2012: Doing Business in a More Transparent World.* Washington, DC: The International Bank for Reconstruction and Development/The World Bank Group, 2011.
Ismail, Ayman, and Rawiah Abdallah. "AUC Venture Lab Business Model." Unpublished internal report. Cairo: American University in Cairo, 2013.
Ismail, Ayman, and Sherif Shabana. "Benchmarking University-based Incubators." Unpublished internal report. Cairo: American University in Cairo, 2013.
Kirby, David. "Entrepreneurship Education: Can Business Schools Meet the Challenge?" *Education + Training* 46 (2004): 510–19.

Masri, Munther, Mohamed Jemni, Ahmed Al-Ghassani, and Aboubakr Badawi. *Entrepreneurship Education in the Arab States.* Beirut: UNESCO, 2010.

NIRAS. *Survey of Entrepreneurship Education in Higher Education in Europe.* Allerød: NIRAS Consultants, 2008.

OECD. *Overcoming Barriers to Administrative Simplification Strategies: Guidance for Policy Makers.* OECD Publishing, 2009.

Saleem, Qamar. *Overcoming Constraints to SME Development in MENA Countries and Enhancing Access to Finance.* Washington, DC: International Finance Corporation/The World Bank Group, 2013.

Schmitt, Christophe. *Entrepreneurship and University: Reflections on the Role of the University Incubators.* Belfort, France: 12th International Conference on Management of Technology, 2003.

UK Universities. *A Forward Look: Highlights of Our Corporate Plan 2001–2004.* London, England: Universities UK, 2000. http://www.universitiesuk.ac.uk/Publications/Documents/CorpPlan2.pdf

Wilson, Karen. *Entrepreneurship Education in Europe.* Brussels: European Foundation for Entrepreneurship Research, 2008.

8. Schumpeterian Entrepreneurs, Total Factor Productivity, and Institutions: Firm-level Data Analysis from Egypt

Karim Badr

Introduction

The quest for growth and development is closely linked to productivity, entrepreneurship, and innovation. According to Joseph Schumpeter,[1] entrepreneurs innovate by introducing new products or services, opening a new market, introducing a new method of production, conquering a new source of raw material, or venturing into a new organization. Through these innovations, new and more efficient methods replace old and obsolete ones in a process of creative destruction, which in turn leads to growth and development.

In a country such as Egypt, with a transitional economy stumbling on the path to development, the entrepreneur has a viable role in spurring economic growth. Economic development in Egypt has been the focus of extensive academic and non-academic studies. The role of entrepreneurs as a linchpin in the process of sustainable growth and development is neglected, however, or as Baumol et al. put it, it is "a missing piece in the puzzle."[2] This chapter aims to explore the role of entrepreneurship in Egypt's economic growth.

Capital, land, and labor drive economic growth through "brute force," while technical change (or productivity) spurs smart growth.[3] Total factor productivity (TFP) is the change in total output that is not attributed to change in the factors of production (land, labor, and capital). Growth in TFP reflects an efficient use of inputs of production (land, labor, and capital). Some scholars attribute the difference in income levels among countries to differences in TFP,[4] or smart growth.[5] TFP is also a measurement of the marginal rate of transformation between inputs and output.

This chapter focuses on the role of innovative entrepreneurs in spurring TFP, the largest source of economic growth in developed countries. It further investigates which economic and institutional settings would spur innovation and entrepreneurship. First, firm-level TFP is estimated, using the World Bank Investment Climate Assessment survey (ICA) panel dataset for the years 2004, 2007, and 2008, after controlling for firm fixed effect to avoid endogeneity. Second, the correlation between TFP and innovation is estimated, in addition to other institutional variables, using the maximum likelihood estimation method. Since innovative firms enjoy higher levels of TFP, as will be shown below, the chapter explores the determinants of innovation at the firm level. The main findings of firm productivity are that innovation is correlated with higher TFP, that firms facing competition are more productive, and that firms located in Upper Egypt have higher productivity levels. Concerning the determinants of innovation, the chapter shows that healthy competition, openness to trade, access to finance, strong governance, and opportunities for employee training are all associated with higher probability of innovation.

In the second part of this chapter, we examine various definitions of the multifaceted concept of entrepreneurship; the relationship between entrepreneurship, economic growth, and TFP; and offer an overview of entrepreneurship in Egypt. Section three includes data description and stylized facts, while section four contains a description of the models used in estimating TFP and assessing the correlation between TFP and innovation, in addition to discussing the results. Since there is a correlation between innovation and higher levels of TFP, we go a step further to inquire about the determinants of innovation. We conclude with summarizing the main findings of the study.

Literature Review
Defining entrepreneurship
"Entrepreneurship is an ill-defined, at best multidimensional, concept."[6]

Although it is not easy to define exactly what an entrepreneur is, shedding light on different definitions in the existing literature helps to understand entrepreneurial characteristics, roles, and problems, in addition to ways of measurement and recommended policy directions. The term entrepreneur was coined by Richard Cantillon[7] to mean a self-employed individual who trades, bakes, or manufactures, and buys a country's products to resell them. Cantillon's definition of an entrepreneur is similar to the prevailing definition of a businessman; it is a vague and broad definition that may apply to innovative or replicative entrepreneurs, productive or rent-seeking entrepreneurs, and opportunity-seeking or necessity entrepreneurs.

During the nineteenth and twentieth centuries, there were various undertakings to define the concept of an entrepreneur more sharply. Alfred Marshall, in his book *Principles of Economics*,[8] focused on the roles of innovator, coordinator, and arbitrageur played by an entrepreneur. Frederick Hawley described an entrepreneur as the owner of output and a risk-taker.[9]

The first main contribution to the concept and theory of entrepreneurship was introduced by Joseph Schumpeter in his 1912 book, *Theory of Economic Development*.[10] Schumpeter emphasized the role of entrepreneur as innovator, an introducer of new goods and new methods of production—someone who opens up new markets, obtains a new source of raw materials, and implements new forms of organization. For Schumpeter, innovation is an idiosyncratic trait of the entrepreneur. Furthermore, he argued that innovative entrepreneurs are the cornerstone of development as they generate 'creative destruction.'[11] He explained that entrepreneurs innovate and are then followed by numerous imitators (replicative entrepreneurs). This process is repeated until new methods wipe away old, obsolete methods, which then results in an economic boom. Ruta Aidis commented that periods of innovation, or lack thereof, are the main drivers of the business cycle.[12] Schumpeter did not add risk-taking to the characteristics of an entrepreneur, in contrast with the common understanding of the term. He clearly stated that an entrepreneur is not a risk-taker, as risk bearing is the primary function of a capitalist. Therefore, the Schumpeterian entrepreneur is an innovator and a source of disequilibria, replacing lower levels of equilibrium with higher and more efficient ones.[13]

Unlike Schumpeter, Knight identified an entrepreneur as one who assumes an uncertainty that will lead him to profit.[14] Edgeworth viewed an entrepreneur as an individual who meets ever-continuing demand—in other words, he maintains a state of equilibrium. Edgeworth claimed that an entrepreneur is a coordinator who combines factors of production and an arbitrageur who links production markets to output markets.[15]

Due to the assumptions underlying these neoclassical theories, the term entrepreneur has carried a connotation of management, as the neoclassical thinkers assume perfect information, rational choice, and perfect certainty. The entrepreneurial role therefore diminished in microeconomic theory.

The significant role that entrepreneurship plays can only be captured if these assumptions were relaxed. "Only by relaxing the assumptions can the entrepreneurial function be included without compromising the model's consistency," according to Aidis.[16] Carree and Thurik presented a more comprehensive definition of entrepreneurship, as

> the manifest ability and willingness of individuals, on their own, in teams, within and outside existing organizations, to perceive and create new economic opportunities (new products, new production methods, new organizational schemes, and new product market combinations) and to introduce their ideas in the market in the face of uncertainty and other obstacles by making decisions on location, form, and the use of resources and institutions.[17]

Nooteboom summarized the various definitions of entrepreneurship as follows: The neoclassical school emphasized the maintenance of equilibrium as the main role of an entrepreneur, while the Austrian school stressed the entrepreneur's ability to combine resources to meet market demands.[18] However, the Schumpeterian school saw the entrepreneur as the creator of disequilibrium and 'creative destruction': "The creation of potential may be seen as Schumpeterian and its realization as Austrian."[19] The Schumpeterian definition of entrepreneurs as causing creative destruction through innovation will be adopted in this research, as the role of innovation is an integral part of the process of development.

Entrepreneurship, TFP, and economic growth

Despite its importance, the role of entrepreneurship in explaining and enhancing TFP and consequently economic growth was totally missed in both neoclassical and endogenous growth models. Although both models underpinned the significance of TFP or technological change in stimulating growth, neither addressed the role of entrepreneurs in fostering TFP. The endogenous growth theory highlights the importance of investment in knowledge, research and development (R&D), and training to spur innovation and accelerate the growth rate. The role of the economic agent behind these investments in knowledge production and innovation was absent, however.

A few efforts were subsequently made to incorporate entrepreneurship into economic growth models. Nonetheless, the ability to pinpoint the relationship between entrepreneurship and economic growth remains elusive or at best indirect. This might be explained by the difficulty associated with measuring entrepreneurship. Several proxies, however, are being used to measure entrepreneurship in terms of a business ownership rate, entry rate, or self-employment rate. None of them are synonyms for entrepreneurship, as business owners might not venture into entrepreneurial activities, while entrepreneurs might not be business owners (intrapreneurs). Scholars have admitted, however, that these proxies served as acceptable measurements for entrepreneurship.[20]

Some authors, though, have reported a negative correlation between economic development and business ownership.[21] When a country enjoys a higher level of per capita income, and consequently higher wages, the opportunity cost of switching from employee to business owner increases. More recent studies have showed different results, however. The rate of business ownership rose in the 1990s and onward, partly due to the emergence of new industries such as information, communication technology, software, biotechnology, and nanotechnology.

Audretsch and Thurik[22] explained that globalization and the information revolution shifted away the comparative advantages of firms in traditional industries. Carree et al.[23] addressed the issue of business ownership rate optimality. They showed that any deviation from the equilibrium (optimal) rate of business ownership will penalize the growth rate of any economy in the medium term. A shortage of business ownership results in decreased competition and efficiency, and hinders innovation. On the other hand, an excessive business ownership rate can cause firms to operate at diseconomies of scale.[24]

Stel and Carree criticized the optimal approach to business ownership for ignoring employment structure per sector.[25] Since rates of business ownership are higher in the service sector compared to manufacturing, deviation from optimal rates may not be a problem of business ownership but rather a problem of sectored structure. They added that sectored business ownership is a function of economic development expected to have a negative relationship with developing countries at the early stages of industrialization. These countries invest heavily in production, distribution, and management to benefit from economies of scale. Meanwhile, at higher levels of economic development, the relationship is expected to be positive, since developed countries have witnessed a shift away from traditional industries toward software, electronics, information and communications technology (ICT), and biotechnology. New technology has moreover undermined the importance of economies of scale. Furthermore, high levels of economic development precede high income levels, which in turn widen the scope of individual demand, consequently creating opportunities for small ventures to appear. Stel and Carree conclude that the relationship between business ownership and economic development is U-shaped, whereby business ownership decreases with low levels of income, then increases as income increases.[26] Stel and Caree demonstrated an equilibrium (optimal) level of business ownership taking into account the sectoral structure, and showed that any deviation from this optimal level harms growth, either because of diseconomies of scale (in cases of too much entrepreneurship), or lack of competition and efficiency (in cases of shortages in entrepreneurs).[27]

Nickell and Nickell et al. showed that competition, measured by number of competitors, has a positive correlation with growth in TFP.[28] Erken et al. used the Compendia dataset, a harmonized dataset for business owners, to assess the role of entrepreneurship in TFP.[29] They re-estimated the econometric models assessing the drivers of productivity presented in five papers and incorporating entrepreneurship to these models measured as the ratio between actual and equilibrium business ownership rates.[30] Their findings confirmed the results of the five seminal studies that public and private research and development, human capital, technology, and other variables are significant drivers of TFP. In addition, they showed that entrepreneurship has a significant impact on TFP.[31]

Audretsch and Keilbach coined the term "entrepreneurial capital," which they defined as "a regional milieu of agents that is conducive to the creation of new firms,"[32] adding that entrepreneurial capital capacity involves dealing with risk and the creation of new business. They showed that there is a strong correlation between regional entrepreneurial capital and regional productivity.[33] Audretsch and Keilbach further argue that while knowledge production is not a sufficient condition for driving growth, entrepreneurs are essential in commercializing knowledge for regions to grow. They proposed that entrepreneurship is an essential vehicle to transform knowledge production into knowledge commercialization. "Thus, ceteris paribus, a greater amount of entrepreneurship is expected to be associated with more diversity and therefore higher growth."[34]

Moreover, the Global Entrepreneurship Monitor concluded that there is a strong relationship between entrepreneurial activities, defined as startup activities, and economic growth.[35] When applied to Egypt, this research emphasizes the role of innovative entrepreneurs as a linchpin of TFP, through the innovation and 'creative destruction' that stimulate growth and development.

Entrepreneurship in Egypt and the required economic settings

While Egypt has pursued economic reforms for the past two decades, entrepreneurial activity in the country remains relatively low. The 2008 Global Entrepreneurship Monitor (GEM) report showed that only 13.1 percent of Egypt's adult population had either tried to start a business or owned and managed their own business.[36] This led to the rank of 11 out of 43 countries surveyed by the GEM, according to the total entrepreneurial activity (TEA) index. Only 6.5 percent of Egyptian TEA enterprises worked in medium or high tech sectors, though, compared to 12.5 percent in Italy (the highest performer in the study). Furthermore, only 9 percent of TEA enterprises in Egypt introduced a new product market combination, which placed Egypt among the lowest four countries.[37]

In this regard, government policies and institutions have been criticized for their weak promotion of entrepreneurial activities. One should therefore ascertain which framework would promote entrepreneurship. More specifically, what are the economic and institutional settings that would accelerate productivity through entrepreneurship? To answer this question, I will discuss the macroeconomic and institutional environment that would help entrepreneurship thrive and would optimally affect productivity and hence economic growth. I will examine several institutional variables. First, the ease of doing business: because making it easy to set up and grow a business encourages an innovative entrepreneur, competition and low market entry and exit barriers are cornerstones of promoting entrepreneurship. Second, facilitating access to finance is essential in encouraging entrepreneurs and in promoting technological innovation and growth.[38] The importance of access to finance for entrepreneurship and growth is well established in the literature, but financing for new and risky businesses is what matters. Third, openness to trade, spending on R&D, optimal taxation systems, and proper corporate governance are all institutional pillars required to promote innovation and productivity growth.

Data Description and Analysis

This chapter makes use of the rich panel dataset of the Investment Climate Enterprise Survey conducted by the World Bank in three-year cycles—2004, 2007, and 2008—for manufacturing firms in Egypt. The dataset is originally composed of 3,129 observations, while the panel dimension applies to only 1,662 observations distributed equally into 554 observations in each year.

Table 8.1. Distribution of firms by economic activity

	2004	2007	2008	Total
Garments	13.90	13.90	11.73	13.18
Textiles	14.08	14.08	17.33	15.16
Machinery and equipment	3.43	3.43	2.35	3.07
Chemicals	13.00	13.00	6.68	10.89
Electronics	4.15	4.15	1.26	3.19
Metal industries	16.06	16.06	18.77	16.97
Non-metal industries	0	0	10.11	3.37
Agro industries	14.08	14.08	1.62	9.93
Other industries	21.30	21.30	30.14	24.25
Total	100.00	100.00	100.00	100.00

Source: Data calculated by the author

Table 8.2. Distribution of firms by size

Number of employees in firm	2004	2007	2008	Total
10 or fewer	12.3	19.5	13.4	15.0
11–50	60.7	51.3	48.7	53.6
51–100	8.5	8.7	10.1	9.1
More than 100	18.6	20.6	27.8	22.3

Source: Data from calculations by the author using Investment Climate Assessment survey for Egypt for 2004, 2007, and 2008

Firms are distributed among sixteen governorates where relative representations remained almost invariant over the three years (see table 8.1.A in the appendix). For a better analysis, these governorates were aggregated into four regions: Cairo; metropolitan areas that include Alexandria, Port Said, and Suez; Lower Egypt (nine governorates); and Upper Egypt (three governorates, including Giza).

Firms are divided into nine main economic activities, as shown in table 8.1.

Firms vary by size in the sample. This chapter classifies firm size according to the number of employees in each firm, where four categories are defined: micro-firms with 10 employees or fewer, small enterprises with between 11 and 50 employees, medium firms with between 51 to 100 employees, and large firms that employ more than 100 employees. The distribution of firms by size is shown in table 8.2. The sample is dominated by small firms with 11 to 50 employees, followed by large firms with more than 100 employees.

Stylized facts

Innovative entrepreneurs are the linchpin of productivity gains. The definition of innovation adopted in this chapter is the introduction of a new product or service to the market, or the upgrading of an existing one. Innovative firms in the manufacturing sector in Egypt hovered around similar rates in each of the three years (see table 8.3). In 2004, 29.6 percent of firms introduced a new product or service, or upgraded an existing one. The rate declined slightly to 27.4 percent in 2007, then rose again to 31 percent in 2008.

Table 8.3. Innovation by year

Firms in manufacturing sector in Egypt	2004	2007	2008	Total
Non-innovative	70.40	72.56	68.95	70.64
Innovative	29.60	27.44	31.05	29.36
Total	100.00	100.00	100.00	100.00

Source: Data from calculations by the author using Investment Climate Assessment survey for Egypt for 2004, 2007, and 2008

Larger firms are more innovative than smaller ones. Table 8.4 shows that as firm size increases, the proportion of those that innovate also increases. Among micro-firms, only 2.4 percent innovate, while the rate increases to 9.4 percent among small firms. Twenty-five percent of medium firms innovate, while 29.3 percent of large firms do. This trend is intuitive, as larger firms have the financial capabilities, the economies of scale, and the expertise needed to innovate and introduce new products and services to the market. It is therefore important to control firm size when measuring the correlation between productivity and innovation, as will later be shown.

Table 8.4. Innovation and firm size

Firm size	Non-innovative	Innovative	Total
Fewer than 10	97.60	2.40	100
11–50	90.56	9.44	100
51–100	74.83	25.17	100
More than 100	70.62	29.38	100

Source: Data from calculations by the author using Investment Climate Assessment survey for Egypt for 2004, 2007, and 2008

Innovation varies according to ownership. The proportion of innovative firms among foreign-owned companies is the highest, at 66.6 percent, followed by domestic companies at 48.4 percent, and bank-owned firms at 42.8 percent (see table 8.5).

Table 8.5. Innovation and ownership

Firm ownership	Non-innovative	Innovative	Total
Individual	73.98	26.02	100
Family	68.92	31.08	100
Domestic company	51.61	48.39	100
Foreign company	33.33	66.67	100
Bank	57.14	42.86	100
Managers of the firm	77.42	22.58	100
Employees of the firm	66.67	33.33	100
Government	63.33	36.67	100
Other	42.86	57.14	100

Source: Data from calculations by the author using Investment Climate Assessment survey for Egypt for 2004, 2007, and 2008

Innovative firms' rates differ substantially from one region to another (see table 8.6). Remarkably, the share of innovative firms in Upper Egypt was the highest in the country at 41 percent, while Lower Egypt came second with 36.2 percent. This result will differ slightly when other variables are controlled, as will be shown in the empirical work. Cairo ranked as the least innovative region (16.4 percent).

Table 8.6. Innovation by region

Region	Non-innovative	Innovative	Total
Cairo	83.54	16.46	100
Metropolitans	77.48	22.52	100
Lower Egypt	63.71	36.29	100
Upper Egypt	58.73	41.27	100

Source: Data from calculations by the author using Investment Climate Assessment survey for Egypt for 2004, 2007, and 2008

Innovation and productivity

Do innovative firms have higher productivity? To answer this question, TFP is estimated using a Cobb-Douglas production function with constant returns to scale, as will be explained below in the model and empirical work. This model indicates that average productivity is higher for innovative firms compared to their non-innovative counterparts (see fig. 8.1).

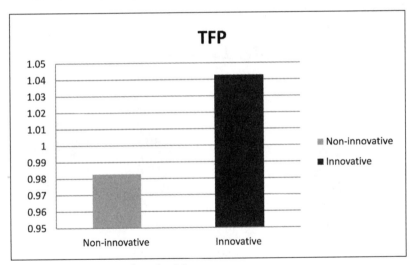

Figure 8.1 Innovation and productivity

Productivity is higher for innovative firms compared to their non-innovative counterparts in many sectors (see fig. 8.2). The difference in productivity between entrepreneurial firms is larger in metal industries, chemicals, garments, and agro business than in other sectors.

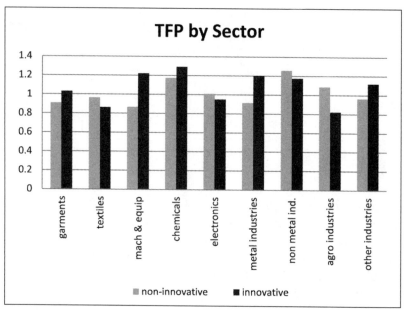

Figure 8.2 Innovation and productivity by sector

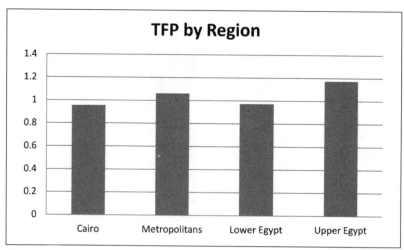

Figure 8.3 Productivity by region

Productivity also varies by region (see fig. 8.3 where TFP is on the y-axis). Interestingly, the region with a relatively higher average TFP (Upper Egypt) is also the region where the rate of innovative firms is higher, as shown in table 8.6.

The Model

This chapter uses a panel data model, with maximum likelihood estimation, to measure the impact of innovation on TFP. Panel data permit determining a causal relationship between growth in TFP and innovation, in addition to another set of investment and institutional climate variables mentioned below. The analysis is done in two stages: estimating TFP, and then using the predicted TFP as a dependent variable with other explanatory variables, as shown below.

TFP is estimated using a log linear Cobb-Douglas production function with constant returns to scale. Following the seminal work of Alvaro Escribano and J. Luis Guasch,[39] the production function is extended to include firm-specific fixed effect to counter the problem of endogeneity that arises from unobserved firm characteristics.

$$Log\ Y_{it} = \beta_0 + \beta_l L_{it} + \beta_m M_{it} + \beta_k K_{it} + \beta_x X_{it} + \beta_{Dr} D_r + \varepsilon$$

Where:
- Y is the firm's output (sales);
- L is the firm's employment;
- M is intermediate materials;
- K is capital stock;
- X is a firm-specific fixed effect vector (firm size, region, and sector);
- D_r is a vector of year dummies;
- ε is the residual (TFP).

All monetary figures are deflated using the Wholesale Price Index to year 2003 prices.

The second equation measures the relationship between TFP and innovation, in addition to a series of other investment climate and institutional variables. Developing a new important product, or updating an existing one, is used as a measure for innovation.

$$TFP_{it} = \beta_0 + \beta_{inn} INN_{it} + \beta_f F_{it} + \beta_{ic} IC_{it} + B_t D_t + \varepsilon$$

Where:
- INN is a dummy for developing a new product or upgrading an existing one;
- F is a vector for firm specific variables such as firm size, region, and sector;

- IC is a vector of variables as follows: competition, corruption, access to finance, percentage of employees with university education or higher, and percentage of skilled employees;
- ε is error term.

After establishing the correlation between TFP and innovation, this chapter looks at the investment and institutional setting that can spur innovation. A logistic regression was run where innovation is the dependent variable, and investment and economic institutional variables are categorized as follows:

Firm-related:
- Research and development;
- Quality of infrastructure;
- Employee training;
- Years of education for management;
- Unionized labor;
- Governance;
- Firm size;
- Firm location.

Market-related:
- Foreign trade;
- Competition;
- Access to finance;
- Ease of doing business;
- Corruption;
- Macroeconomic uncertainty;
- Perception of legal issues as a problem;
- Perception of tax rate.

Results
The estimation of total factor productivity is shown in table 8.7. Two different regressions were run to estimate the correlation between TFP and innovation in addition to other institutional variables. The two methods are maximum likelihood estimation and ordinary least squares (OLS) estimation. The results of the two methods are almost identical, which enforces the consistency of the correlation between the variables (see table 8.8).

There is a significant relationship between innovation and TFP. Innovative firms have a productivity premium of 9 percent compared to non-innovative firms. The result is robust after controlling for firm-specific fixed effects to avoid endogeneity, and controlling for other variables.

Interestingly, firms that face competition are more productive than non-competing firms. This signifies the importance of leveling the playing field by making it easier to do business, and keeping low barriers to market entry and exit. Employee education is positively correlated with TFP. A higher proportion of employees with at least a university education is associated with higher productivity (14 percent premium). One counterintuitive result is that access to finance is associated with lower productivity. Firms that carry loans have lower productivity than their counterparts who do not have loans.

Firms located in Upper Egypt are more productive than their counterparts located in Cairo, and firms located in metropolitan areas have productivity that is not significantly different from those in Cairo. This raises a question concerning the relationship between business clustering, economies of agglomeration, and productivity that is beyond the scope of this chapter.

Determinants of Innovation

Since innovative entrepreneurship is correlated with TFP, it is important to study the economic variables and institutions that spur innovation. To further explore the determinants of innovation, and to have a better insight into the institutional variables that could promote it, a logistic regression was run (see table 8.9). The results concerning firm-related variables show that R&D is as essential to innovation; having an R&D department in a firm increases the probability of introducing a new product by 73 percent. Providing training to employees enhances the chances of innovation by 34 percent. Firms that are open to trade are more innovative than their counterparts that are not. Firms with an external auditor are more innovative. Firms located in Upper Egypt, once again, are not only more productive than firms in Cairo or in metropolitan areas but also are more innovative. Table 8.4 shows that the probability of being innovative increases with firm size.

As for market-related factors, competition is not only correlated with higher productivity but also drives innovation (38 percent premium). Firms that face competition are more innovative through introducing new goods and services. Competition might force firms to innovate to increase profitability and market share, and it indirectly helps productivity. Once again, this result underscores the importance of free markets, lowering barriers to entry, and enforcing free competition.

Foreign trade is significantly and positively correlated with higher innovation. This result further supports the importance of openness to trade and free market competition to foster innovation.

Access to finance, although negatively correlated with productivity, is positively correlated with innovation. As a firm's sources of finance increase, the opportunity for innovation expands. Taxation, or the perception of a high tax rate, is negatively correlated with innovation. Lowering taxes might act as an incentive to promote innovation, though using a perception index might be problematic.

Conclusion

Schumpeter outstandingly recognized the role of the entrepreneur as an innovator in the process of development. Innovations displace the existing equilibrium and replace it with a higher one. Schumpeter emphasized that the introduction of a new product or service, among other factors, would result in 'creative destruction' as new ways sweep away old, obsolete methods.

This chapter highlights the relationship between innovative firms and productivity, where the latter accounts for the largest share of growth. It shows that innovative firms enjoy higher levels of productivity, after controlling for firm fixed effects. It also highlights the significance of openness to trade, competition, and ease of doing business in promoting innovation and consequently productivity. It stresses the importance of research and development, education, governance, and access to finance for both productivity and innovation.

It is evident that adopting free market policies, reducing customs, removing barriers to market entry and exit, and enforcing competition are important pillars in building the appropriate institutions to promote entrepreneurship. Egypt is still suffering from high barriers to market entry, represented in cumbersome registration and licensing processes, red tape, low contract enforcement, and low access to finance. Providing easier access to finance, enforcing corporate governance, and linking research and development with the business community are crucial to promote productivity and innovation.

Table 8.7. TFP estimation

Variables	Log deflated sales
Log deflated intermediate material	0.614 (0.0153)***
Log deflated capital	0.0966 (0.0131)***
Log number of employees	0.250*** (0.0433)
Firm size 11–50	0.0373 (0.0735)
Firm size 51–100	0.257** (0.127)
Firm size 100+	0.288* (0.162)
Metropolitans	0.121 (0.0769)
Lower Egypt	0.0428 (0.0594)
Upper Egypt	0.188** (0.0837)
Textiles	0.0272 (0.0910)
Machinery and equipment	0.0422 (0.150)
Chemicals	0.292*** (0.0974)
Electronics	0.110 (0.147)
Metal	0.0591 (0.0860)
Non-metals	0.151 (0.148)
Agro industries	0.137 (0.103)
Other industries	0.0546 (0.0811)
d07	-0.0640 (0.0559)
d08	0.161*** (0.0589)
Constant	1.304*** (0.127)
Observations	1,561
Number of unitid	554

Standard errors in parentheses.   *** p<0.01, ** p<0.05, * p<0.1

Table 8.8. Innovation and TFP by two methods: OLS and MLE

Variables	OLS TFP	MLE TFP	(2) sigma_u	(3) sigma_e
Innovation	0.0921* (0.0558)	0.0919* (0.0554)		
Competition	0.191*** (0.0741)	0.193*** (0.0736)		
Corruption	-0.0781 (0.0577)	-0.0780 (0.0573)		
Borrowing	-0.109* (0.0639)	-0.109* (0.0635)		
% employees with university education +	0.140** (0.0652)	0.141** (0.0647)		
% of skilled employees	-0.0173 (0.0181)	-0.0171 (0.0179)		
Firm size 11–50	-0.0469 (0.0684)	-0.0458 (0.0680)		
Firm size 51–100	-0.000797 (0.1000)	-0.00206 (0.0995)		
Firm size 100+	-0.133 (0.0859)	-0.130 (0.0857)		
Metropolitans	0.112 (0.0780)	0.112 (0.0782)		
Lower Egypt	-0.000451 (0.0595)	-0.000671 (0.0596)		
Upper Egypt	0.164* (0.0844)	0.163* (0.0847)		
Textiles	-0.00245 (0.0908)	-0.00315 (0.0908)		
Machinery and equipment	0.0162 (0.150)	0.0175 (0.150)		
Chemicals	0.251*** (0.0966)	0.250*** (0.0966)		
Electronics	0.0219 (0.147)	0.0232 (0.147)		
Metal	0.0533 (0.0854)	0.0542 (0.0855)		
Non-metals	0.216 (0.145)	0.217 (0.145)		
Agro industries	0.0606 (0.101)	0.0620 (0.101)		
Other industries	0.0584 (0.0803)	0.0595 (0.0804)		
Constant	-0.0699 (0.0857)	-0.0712 (0.0857)	0.142* (0.0760)	0.884*** (0.0197)
Observations	1,554	1,554	1,554	1,554
Number of unitid	554	554	554	554

Table 8.9. Determinants of innovation

Variables	Innovation	lnsig2u
R&D	0.733*** (0.178)	
d07	0.0235 (0.161)	
d08	-0.0680 (0.188)	
Competition	0.388* (0.235)	
Training	0.341* (0.182)	
Managers' years of education	0.0220 (0.0226)	
Foreign trade	0.0121*** (0.00362)	
Borrowing	0.616*** (0.167)	
Governance	0.481** (0.187)	
Macroeconomic uncertainty	-0.125 (0.194)	
Infrastructure	0.195 (0.136)	
Business license	0.178 (0.149)	
Legal problems	0.126 (0.148)	
Metropolitans	0.225 (0.251)	
Lower Egypt	0.923*** (0.183)	
Upper Egypt	1.115*** (0.245)	
Tax rate	-0.281* (0.155)	
Corruption	0.750*** (0.161)	
Firm size 11–50	0.574** (0.258)	
Firm size 51–100	1.327*** (0.319)	
Firm size 100+	1.271*** (0.301)	
Constant	-3.681*** (0.476)	-1.080* (0.575)
Observations	1,657	1,657
Number of unitid	554	554

Standard errors in parentheses. *** $p<0.01$, ** $p<0.05$, * $p<0.1$

Appendix

Table 8.1.A. Distribution of firms by governorate

Governorate	2004	2007	2008
Cairo	28.52	28.52	28.52
Alexandria	11.55	11.55	11.55
Port Said	1.62	1.62	1.26
Suez	0.18	0.18	0.54
Damietta	1.26	1.26	0.9
Dakahliya	4.69	4.69	4.87
Sharkiya	12.64	12.64	12.64
Qualyubia	10.83	10.83	11.01
Kafr al-Sheikh	0.90	0.90	0.72
Gharbiya	8.12	8.12	8.12
Menoufiya	5.6	5.6	5.78
Beheira	1.99	1.99	1.99
Ismailia	0.72	0.72	0.72
Giza	8.84	8.84	8.84
Beni-Suef	0.90	0.90	0.90
Minya	1.62	1.62	1.62
Total	100.00	100.00	100.00

Notes

1 Joseph Schumpeter, *Capitalism, Socialism and Democracy* (London: Routledge, 1942).
2 William J. Baumol, Robert E. Litan, and Carl J. Schramm, *Good Capitalism, Bad Capitalism, and the Economics of Growth and Prosperity* (New Haven: Yale University Press, 2007).
3 Baumol, Litan, and Schramm, *Good Capitalism*.
4 William Easterly and Ross Levine, "What Have We Learned from a Decade of Empirical Research on Growth? It's Not Factor Accumulation: Stylized Facts and Growth Models," *World Bank Economic Review* 15 (2001): 177–219; Peter Joseph Klenow and Andrés Rodriguez-Clare, "The Neoclassical Revival in Growth Economics: Has It Gone Too Far?" *NBER Macroeconomics Annual 1997* 12 (1997): 73–10.
5 Baumol et al. Litan, and Schramm, *Good Capitalism*.
6 Roy A. Thurik and Sander Wennekers, "Linking Entrepreneurship and Economic Growth: Small Business Economics," *Small Business Economics* 13 (1999): 27–56.
7 Richard Cantillon, *Essai sur la Nature du Commerce en Général* (Piscataway, NJ: Transaction Publishers (English edition), 1755 [2001]).

8   Alfred Marshall, *Principles of Economics* (London and New York: MacMillan & Co., 1980).

9   Frederick B. Hawley, "The Risk Theory of Profit," *Quarterly Journal of Economics* 7 (1893): 459–79.

10 Joseph Schumpeter, *The Theory of Economic Development* (Cambridge: Harvard University Press, 1912).

11 Joseph Schumpeter, *Capitalism, Socialism and Democracy*, 82–83.

12 Ruta Aidis, "Entrepreneurship in Transition Countries: A Review," Working Papers 61, Center for the Study of Economic and Social Change in Europe, University College London (2005).

13 Schumpeter, *The Theory of Economic Development*.

14 Frank Knight, *Risk, Uncertainty and Profit* (Chicago: University of Chicago Press, 1912).

15 Francis Edgeworth, *Paper Relating to Political Economy* (London: Macmillan and Co. Ltd., 1925).

16 Ruta Aidis, "Entrepreneurship and Economic Transition," Tinbergen Institute Research Paper (2003), 5.

17 Martin Carree and Roy A. Thurik, *The Impact of Entrepreneurship on Economic Growth: Handbook of Entrepreneurship Research* (Boston: Dordrecht, 2003), 441.

18 Bart Nooteboom, "Schumpeterian and Austrian Entrepreneurship: A Unified Process of Innovation and Diffusion," Research Report no. 1993–01 (Groningen: Groningen University, 1993).

19 Nooteboom, "Schumpeterian and Austrian Entrepreneurship," 1.

20 André Van Stel and Martin Carree, *Business Ownership and Sectoral Growth: Scientific Analysis of Entrepreneurship and SMEs* (Zoetermeer: Scales Research Reports, 2002).

21 Simon Kuznets, *Economic Growth of Nations, Total Output and Production Structure* (Cambridge, MA: Harvard University Press and Belknapp Press, 1971); Paul T. Schultz, "Women's Changing Participation in the Labor Force: A World Perspective," Policy Research Working Paper Series 272, The World Bank, 1989; Gustavo Yamada, "Urban Informal Employment and Self-Employment in Developing Countries: Theory and Evidence," *Economic Development and Cultural Change* 44 (1996): 289–314.

22 David Audretsch and Roy Thurik, "Impeded Industrial Restructuring: The Growth Penalty," C.E.P.R. Discussion Papers 2648 (London: Centre of Economic Policy Research, 2000).

23 Martin A. Carree et al., "Economic Development and Business Ownership: An Analysis Using Data of 23 OECD Countries in the Period 1976–1996," *Small Business Economics* 19 (2002): 271–90.

24 Martin A. Carree et al., "Economic Development and Business Ownership."

25 Van Stel and Carree, *Business Ownership and Sectoral Growth*.

26 Van Stel and Carree, *Business Ownership and Sectoral Growth*.

27 Van Stel and Carree, *Business Ownership and Sectoral Growth*.

28 Stephen J. Nickell, "Competition and Corporate Performance," *Journal of Political Economy* 104 (1996): 724–46; Stephen J. Nickell, Daphne Nicolitsas, and Neil Dryden, "What Makes Firms Perform Well?" *European Economic Review* 41 (1997): 783–96.

29 Hugo Erken, Piet Donselaar, and Roy Thurik, *Total Factor Productivity and the Role of Entrepreneurship*, Tinbergen Institute Discussion Paper, Netherland. Tinbergen Institute. 2008.

30 Nicolas Belorgey, Remy Lecat, and Tristan-Pierry Maury, "Determinants of Productivity per Employee: An Empirical Estimation Using Panel Data," *Economics Letters* 91 (2006): 153–57; David Theodore Coe and Elhanan Helpman, "International R&D Spillovers," *European Economic Review* 39 (1995): 859–87; Hans-Jürgen Engelbrecht, "International R&D Spillovers, Human Capital and Productivity in OECD Economies: An Empirical Investigation," *European Economic Review* 41 (1997), 1479–88; Rachel Griffith, Stephen Redding, and John Van Reenen, "Mapping the Two Faces of R&D: Productivity Growth in a Panel of OECD Industries," *Review of Economics and Statistics* 86 (2004): 883–95; Dominique Guellec and Bruno Van Pottelsberghe de la Potterie, "From R&D to Productivity Growth: Do the Institutional Settings and the Source of Funds of R&D Matter?" *Oxford Bulletin of Economics and Statistics* 66 (2004): 353–78.
31 Erken, Donselaar, and Thurik, *Total Factor Productivity.*
32 David Audretsch and Max Keilbach, "Does Entrepreneurship Capital Matter?" *Entrepreneurship: Theory & Practice* 28 (2004): 420.
33 Audretsch and Keilbach, "Does Entrepreneurship Capital Matter?"
34 Audretsch and Keilbach, "Does Entrepreneurship Capital Matter?"
35 Global Entrepreneurship Monitor (GEM), 2000.
36 Hala Hattab, *Global Entrepreneurship Monitor: Egypt Entrepreneurship Report, 2008* (Cairo: The Industrial Modernization Centre, 2008).
37 Hattab, *Global Entrepreneurship Monitor, 2008.*
38 Schumpeter, *The Theory of Economic Development*
39 Alvaro Escribano and Luis J. Guasch, *Assessing the Impact of the Investment Climate on Productivity Using Firm-level Data: Methodology and the Cases of Guatemala, Honduras, and Nicaragua* (Washington D.C., World Bank Policy Research Working Paper Series 3621, 2005).

Bibliography

Aidis, Ruta. "Entrepreneurship and Economic Transition." Tinbergen Institute Research Paper. Netherland. Tinbergen Insititute. 2003.

———. "Entrepreneurship in Transition Countries: A Review." Working Papers 61, Center for the Study of Economic and Social Change in Europe, University College London, 2005. Center for the Study of Economic and Social Change in Europe.

Audretsch, David, and Max Keilbach. "Does Entrepreneurship Capital Matter?" *Entrepreneurship: Theory & Practice* 28 (2004): 419–29.

Audretsch, David, and Roy Thurik. "Impeded Industrial Restructuring: The Growth Penalty." C.E.P.R. Discussion Papers 2648. London: Centre of Economic Policy Research, 2000.

Baumol, William. J., Robert E. Litan, and Carl J. Schramm. *Good Capitalism, Bad Capitalism, and the Economics of Growth and Prosperity.* New Haven: Yale University Press, 2007.

Belorgey, Nicolas, Remy Lecat, and Tristan-Pierry Maury. "Determinants of Productivity per Employee: An Empirical Estimation Using Panel Data." *Economics Letters* 91 (2006): 153–57.

Cantillon, Richard. *Essai sur la Nature du Commerce en Général.* Piscataway, NJ: Transaction Publishers (English edition), 1755 [2001].

Carree, Martin, and Roy A. Thurik. *The Impact of Entrepreneurship on Economic Growth: Handbook of Entrepreneurship Research.* Boston: Dordrecht, 2003.

Carree, Martin, André J. Van Stel, Roy A. Thurik, and Sander Wennekers. "Economic Development and Business Ownership: An Analysis Using Data of 23 OECD Countries in the Period 1976–1996." *Small Business Economics* 19 (2002): 271–90.

Coe, David Theodore, and Elhanan Helpman. "International R&D Spillovers." *European Economic Review* 39 (1995): 859–87.

Easterly, William, and Ross Levine. "What Have We Learned from a Decade of Empirical Research on Growth? It's Not Factor Accumulation: Stylized Facts and Growth Models." *World Bank Economic Review* 15 (2001): 177–219.

Edgeworth, Francis. *Paper Relating to Political Economy.* London: Macmillan and Co. Ltd., 1925.

Engelbrecht, Hans-Jürgen. "International R&D Spillovers, Human Capital and Productivity in OECD Economies: An Empirical Investigation." *European Economic Review* 41 (1997): 1479–88.

Erken, Hugo, Piet Donselaar, and Roy Thurik. *Total Factor Productivity and the Role of Entrepreneurship.* Tinbergen Institute Discussion Paper, 2008.

Escribano, Alvaro, and Luis J. Guasch. "Assessing the Impact of the Investment Climate on Productivity Using Firm-Level Data: Methodology and the Cases of Guatemala, Honduras, and Nicaragua." World Bank Policy Research Working Paper Series 3621, 2005. Washington D.C. World Bank.

The Global Entrepreneurship Monitor (GEM) Executive Report. London. London Business School. 2000.

Griffith, Rachel, Stephen Redding, and John Van Reenen. "Mapping the Two Faces of R&D: Productivity Growth in a Panel of OECD Industries." *Review of Economics and Statistics* 86 (2004): 883–95.

Guellec, Dominique, and Bruno Van Pottelsberghe de la Potterie. "From R&D to Productivity Growth: Do the Institutional Settings and the Source of Funds of R&D Matter?" *Oxford Bulletin of Economics and Statistics* 66 (2004): 353–78.

Hattab, Hala. *Global Entrepreneurship Monitor: Egypt Entrepreneurship Report, 2008.* Cairo: The Industrial Modernization Centre, 2008.

Hawley, Frederick B. "The Risk Theory of Profit." *Quarterly Journal of Economics* 7 (1893): 459–79.

Klenow, Peter Joseph, and Andrés Rodriguez-Clare. "The Neoclassical Revival in Growth Economics: Has It Gone Too Far?" *NBER Macroeconomics Annual 1997* 12 (1997): 73–102.

Knight, Frank. *Risk, Uncertainty and Profit.* Chicago: University of Chicago Press, 1912.

Kuznets, Simon. *Economic Growth of Nations, Total Output and Production Structure.* Cambridge, MA: Harvard University Press and Belknapp Press, 1971.

Marshall, Alfred. *Principles of Economics.* London and New York: MacMillan & Co., 1980.

Nickell, Stephen J., "Competition and Corporate Performance." *Journal of Political Economy* 104 (1996): 724–46.

Nickell, Stephen J.. Daphne Nicolitsas, and Neil Dryden. "What Makes Firms Perform Well?" *European Economic Review* 41 (1997): 783–96.

Nooteboom, Bart. "Schumpeterian and Austrian Entrepreneurship: A Unified Process of Innovation and Diffusion." Research report no. 1993–01. Groningen: Groningen University, 1993.

Schultz, Paul T. "Women's Changing Participation in the Labor Force: A World Perspective." Policy Research Working Paper Series 272. Washington D.C. The World Bank, 1989.

Schumpeter, Joseph. *Capitalism, Socialism and Democracy*. United Kingdom: Routledge, 1942.

———. *The Theory of Economic Development*. Cambridge: Harvard University Press, 1912.

Thurik, Roy A., and Sander Wennekers. "Linking Entrepreneurship and Economic Growth: Small Business Economics." *Small Business Economics* 13 (1999): 27–56.

Van Stel, André, and Martin Carree. *Business Ownership and Sectoral Growth: Scientific Analysis of Entrepreneurship and SMEs*. Zoetermeer: Scales Research Reports, 2002.

Van Stel, André, Martin Carree, and R. Thurik. "The Effect of Entrepreneurial Activity on National Economic Growth." *Small Business Economics* 24 (2005): 311–21.

World Bank Investment Climate Assessment. Washington D.C.: The World Bank. 2004, 2007, and 2008.

Yamada, Gustavo. "Urban Informal Employment and Self-employment in Developing Countries: Theory and Evidence." *Economic Development and Cultural Change* 44 (1996): 289–314.

9. The Egyptian Revolution: An Entrepreneurial Act? A Personal Account

Khaled Ismail

On January 25, 2011, Egyptians swarmed the streets in protest. On that day, I was sitting in Intel's headquarters in Santa Clara, getting ready to conduct the final round of negotiations to sell the company I had founded in Egypt back in 2002.[1] While waiting for my hosts in the large meeting room, my mind wandered back to the time I founded my first startup in 1991, and the moments that followed. I found it amazing that twenty years of being an entrepreneur could be summarized in a handful of highlights and another handful of shortcomings. It was a twenty-second summary of a twenty-year journey that was only interrupted when my hosts arrived in the meeting room.

They had obviously heard more than I had about what was happening in Cairo on that day, and politely explained to me that we needed to postpone the negotiations—and probably the entire deal. I went back to where I was staying, turned on the television, and was mesmerized by the scene. I kept switching between channels for an entire week until I managed to get on the first flight back to Cairo.

The day after I arrived, I walked from where I live to Tahrir Square. I wanted to get a feel of what I had been watching on television for the past week. I wanted to interact with the people on the streets and find an answer to a very important question: Are these people on the streets entrepreneurial in what they are doing? Is there an analogy between the behavior of an entrepreneur founding a startup and that of the Egyptian people throughout the revolution? I have been confronting this question for the past four years, unable to reach a final verdict.

On March 13, 2011, I signed the agreement with Intel, and the 120 engineers working at my startup, SySDSoft, became Intel engineers. This news made it around the world, as SySDSoft was the global leader in providing the protocol stack (the software brain) of next-generation mobile devices, while Intel is one of the largest technology companies in the world. It marked the first time that a global technology product was developed by someone in the Middle East and Africa, excluding Israel.

Since signing the deal, I have worked with hundreds of young entrepreneurs in Egypt, never forgetting the question that crossed my mind about the Egyptian revolution. Every time I meet with a new startup, I come back to the same question about the analogy between that startup and the revolution. I found myself referring to the revolution more and more when I was warning young entrepreneurs about the pitfalls that plague startups. Finally, I decided to put it all together in writing this piece.

How Did It Start?

The number of Egyptians dissatisfied with former President Hosni Mubarak's regime had been mounting since 2005, when the parliamentary elections were clearly rigged, Mubarak decided to run for president for a fifth term, and the regime started preparing his son Gamal to inherit the presidency. The poor, in particular, were feeling mounting economic pressure, which was to a large extent the result of corruption.

In entrepreneurial terms, this is called the 'pain.' Excellent entrepreneurial ideas are often rooted in a desire to solve an existing problem, or remove a pain that the end-user is suffering from. The number of Egyptians feeling the pain was growing, as evident from the demonstrations that started occurring in 2008—but they were limited in size and reach, and were always combated by the police and the secret service.

As seen in figure 9.1, the flat portion of the curve corresponds to the period from 2008 to early 2011 in Egypt. During the initial flat portion of the entrepreneurial S-curve, entrepreneurs are supposed to think about the details of their project and its chances of success, and to put together a corresponding business plan. The more time entrepreneurs take in creating their business plan, the better prepared and equipped they will be for their entrepreneurial endeavor. There is no prescribed ideal time for this initial stage, though. Contrary to the length of a runway for an airplane taking off, the "runway" of a startup can vary significantly, from just a few months to several years.

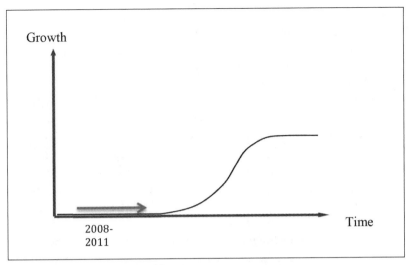

Figure 9.1 The entrepreneurial S-curve

Source: Whitney Johnson. "Throw Your Life a Curve," 3 September, 2009
http://blogs.hbr.org/2012/09/throw-your-life-a-curve/

In the case of the January 2011 revolution, the question is whether January 2011 was the right moment for takeoff. Was the three-year "incubation" period ahead of the revolution sufficient for proper preparation? Had the "founders" of the revolution done their homework and put together the right "business plan?" Had they conducted "what-if" scenarios in case their original plan failed? Had they carefully studied the landscape and the competition? All of these questions are exactly the same ones we ask startups before deciding to invest in them, and can be applied to a revolution by analogy.

Execution

The demonstrations were planned to be peaceful, and to a large extent remained such. The demonstrators, unsure of the outcome, were gauging their next steps based on the "market" response. When the number of additional demonstrators started climbing faster than any of the "founders" had expected, the "barrier to entry" was lowered significantly. Based on that better-than-expected "market acceptance," the demonstrators raised the bar of their requests day by day until they managed to oust Mubarak on February 11, 2011—by all measures, an extremely fast collapse for a thirty-year-old regime.

On February 11, 2011, from 6 p.m. until late into the night, Tahrir Square was transformed into a celebration site. I was there to share the celebration with the millions on the streets and in the square, but I could not resist asking the difficult questions. I wondered, what are we celebrating and why are we so happy?

The next morning, people went to clean Tahrir Square, basically implying that the job was done. In startups, there is a well-known phenomenon called "early celebration," which happens when a startup gets its first customer or its first income. That point marks the first upward bend in the S-Curve, similar to the airplane leaving the ground. One major problem associated with early celebration is that the founders do not realize they have only barely taken off, and still have a very long road to achieving their ultimate goals.

Quite often, startup founders do not pay much attention to documenting the shareholding structure or the decision-making process during the incubation phase. During that phase, when all are unified toward a short-term goal of just wanting to take off, the lack of these details may not matter. Following an early celebration once some success is achieved, however, heated debates often erupt around common and crucial questions, such as the shareholding structure and the decision-making process.

The January 2011 revolution never managed to answer these questions in the aftermath of Mubarak's ouster. Many stakeholders claimed ownership, ranging from the military to the Islamists to the socialists and leftists, all the way to the soccer supporters commonly known as the Ultras. When conflict starts, the shareholder with the largest stake dominates, and typically the minority shareholders get crushed. The Muslim Brotherhood appeared as the largest shareholder and decided to drive the revolution. In startups, at such a turning point the company changes direction as the new management starts changing the course. Silent shareholders wait to see whether that new change will bring success to the company.

Those who started the revolution and claimed success on February 11, 2011, did clearly undermine the "market forces." Typically, a market includes different players competing for a larger market share. On the one hand, the "customer" (the Egyptian people) was now poised to see change for the better in terms of living standards, services, and freedom. On the other hand, the "competition"—in the form of the *feloul* (remnants of the ousted regime) and the deep, corrupt state—would not sit and watch its "market share" evaporate without trying to counterattack. Thus, the revolution found itself torn between new management and fierce competition, while it began failing to deliver its promise to the customer.

The new management, the Brotherhood, made the same mistake that new management so often does: it ignored the original vision and mission, did not pay much attention to the customers, and undermined the competition, thinking that it could create success despite these very complex boundary conditions. It is not uncommon for such new management to fail in a relatively short period of time, which is exactly what happened. There is an enormous difference between pivoting around the original idea while maintaining focus on the customers and the market and starting up an entire new venture catering to a different customer base. The former only rectifies the path, capitalizing on already created values, while the latter is equivalent to starting from scratch.

Can the 'Company' Be Saved?

This is always a tough question to ask when a company is suffering after failing to deliver to its customers. Typically, there are only two solutions to the problem: go back to the drawing board and draft a new business plan that makes more sense, or start a new company where you can possibly address these shortcomings and apply all the lessons learned from the previous trial.

Interestingly enough, both solutions were being applied simultaneously to the Egyptian revolution. Some forces were trying to redefine the course to achieve the original goals, while others were working on a new startup (Revolution 3.0). As both were catering to the same customer and tackling the same problem, obviously only one of them would succeed. The winner would be the one who understood the lessons learned more than the other, and identified clear solutions to the shortcomings that had prevented success thus far.

This extremely brief analogy between the Egyptian revolution and the life of a startup may sound simplistic. Nevertheless, the similarities cannot be ignored. It is not a surprise that in the same way that most startups fail, statistically most revolutions fail as well. History celebrates only the very successful revolutions, much as only the most prominent startups which achieve vast success become globally recognized. In both cases, failure comes with an upside, which is an enriched experience that could not have been achieved otherwise. One can go to history books to read about all the revolutions that have taken place, and similarly one can go and read books and books about entrepreneurship and startups and guidelines for success. But none of that can replace personal experience.

The Startup Landscape in Egypt after January 25

Following the early celebration of the Egyptian revolution, a sense of courage prevailed among the youth. They were finally allowed to dream. Courage and risk-taking are mistakenly interpreted as cultural issues, suggesting that some cultures are less courageous and more risk-averse than others. The fact is that these characteristics are more affected by environment than by culture. Wherever the environment rewards risk-takers and encourages people to think critically, we find more entrepreneurs. Thus, with the change of environment in Egypt after February 11, 2011, we witnessed a flood of thousands of startups. Just as the revolution did not learn from its analogy with startups, though, the new entrepreneurs did not learn from all the mistakes they experienced firsthand in the revolution.

Many startups identified an interesting idea that promised value to their customers, similar to the revolution's grand promises to the Egyptian people. Yet a good idea is rarely sufficient for success. In fact, ideas are a dime a dozen. Business plan competitions are now frequent events, and many startups saw winning these competitions as a goal in itself. Many evaluations during these competitions focus on the idea rather than the details of the plan and the underlying assumptions.

As most of these entrepreneurs were still quite young and inexperienced (similar to the majority of those who showed up in Tahrir Square on January 25), they clearly lacked an in-depth understanding of market conditions, such as market segmentation, barriers to entry, competition, pricing, and so on. Hand-waving assumptions only lead to misleading business plans and incorrect financials. It is thus no surprise that most of the startups inaugurated in 2011 that were lucky enough to raise some money are now running out of funds before achieving their promised plans. As unfortunate as this may sound, it is a price that has to be paid to compensate for the lack of experience.

Mentorship can play a crucial role in shortening the period of the learning curve, or in avoiding a startup's complete failure. Statistically, those entrepreneurs who are mentored by more experienced entrepreneurs, and who are willing to listen to their mentors, stand a much better chance of success. A mentor should not be mistaken for a consultant, advisor, or a board member. A mentor is there to raise good questions rather than identify solutions. A mentor gives guidelines and raises flags rather than giving operational instructions. Successful mentors are characterized by their long experience as serial entrepreneurs, as well as by strong interpersonal skills. One characteristic cannot compensate for the other, and unfortunately the availability of mentors who are strong in both is quite rare. There is no lack of mentors with good intentions who are willing to give, but they need to climb their own learning curve as the overall entrepreneurship scene is being developed.

The good news is that several years after the revolution, a good portion of the bumpy path is now behind us. There is evidence that some of the two- to three-year-old startups still stand a good chance of success. A much wider ecosystem has developed today, ranging from angel investors to venture capital, from NGOs to academic and governmental institutions, all willing to contribute positively to the growth of entrepreneurship.

Yet to Come

Whether there will be a happy ending is something for history to judge. Nevertheless, the political changes in Egypt and the evolution of the entrepreneurial ecosystem will continue to go hand in hand. One of them will lead and help the other, and it is not necessarily the political system that will take the lead. It is conceivable that new political leaders will evolve out of the entrepreneurship scene, which is characterized by risk-takers who think "out of the box." Should more and more entrepreneurs succeed in building mature small and medium enterprises, this will drive the economy in a way that has not happened in the past sixty years, a period during which the government has always been the primary driver of the economy. Thus, successful entrepreneurship may actually lessen the government's role and force economic decentralization.

Of course, the government's political and socioeconomic direction will have an impact on entrepreneurs. The regime may feel threatened by the young entrepreneurs, and hence may complicate the regulatory scene to maintain its control. On the other hand, the regime could be wise enough to realize that the Egyptian economy will not grow without the contribution of the youth. The ruling regime's decisions in the coming two years will have an enormous impact on Egypt's future.

Notes
1   Khaled Ismail first drafted this personal account in December 2013, based on a dinner talk he gave to the 2012 AUC Research Conference guests.

Bibliography
Whitney Johnson. "Throw Your Life a Curve," September 3, 2009. https://hbr.org/2012/09/throw-your-life-a-curve/